THE THEORY OF FINITE LINEAR SPACES

This is the first comprehensive text to cover finite linear spaces. It contains all the important results that have been published up to the present day, and is designed to be used not only as a resource for researchers in this and related areas, but also as a graduate level text. In eight chapters the authors introduce and review fundamental results, and go on to cover the major areas of interest in linear spaces. A combinatorial approach is used for the greater part of the book, but in the final chapter recent advances in group theory relating to finite linear spaces are presented. At the end of each chapter there is a set of exercises which are designed to test comprehension of the material, and there is also a section of research problems. It will be an invaluable book for researchers in geometry and combinatorics as well as an excellent text for graduate students.

THE THEORY OF FINITE LINEAR SPACES

COMBINATORICS OF POINTS AND LINES

LYNN MARGARET BATTEN
University of Manitoba, Canada

ALBRECHT BEUTELSPACHER
Justus-Liebig Universität, Germany

CAMBRIDGE UNIVERSITY PRESS
Cambridge, New York, Melbourne, Madrid, Cape Town, Singapore, São Paulo, Delhi

Cambridge University Press
The Edinburgh Building, Cambridge CB2 8RU, UK

Published in the United States of America by Cambridge University Press, New York

www.cambridge.org
Information on this title: www.cambridge.org/9780521114189

First published 1993
This digitally printed version 2009

A catalogue record for this publication is available from the British Library

Library of Congress Cataloguing in Publication data

Batten, Lynn Margaret.

The theory of finite linear spaces / Lynn Margaret Batten,
Albrecht Beutelspacher.

p. cm.

Includes bibliographical references and index.
ISBN 0-521-33317-2

1. Vector spaces. I. Beutelspacher, Albrecht. II. Title.
III. Title: Finite linear spaces.
QA186.B38 1993
512′.52–dc20 92-45860
CIP

ISBN 978-0-521-33317-7 hardback
ISBN 978-0-521-11418-9 paperback

Contents

	Preface	*page* ix
1	**The essentials**	1
1.1	Definitions and examples	1
1.2	Affine spaces and projective spaces	3
1.3	The connection between affine and projective spaces	10
1.4	The history of finite linear spaces	11
1.5	Some basic results and the Fundamental Theorem	14
1.6	Designs	17
1.7	Another example of a linear space	18
1.8	Exercises	19
1.9	Research problems	20
1.10	References	20
2	**Complementation**	22
2.1	Aim of the chapter	22
2.2	The details	23
2.3	The pseudo-complement theorems	25
2.4	Exercises	31
2.5	Research problems	31
2.6	References	32
3	**Line sizes**	33
3.1	Introduction	33
3.2	Two consecutive line sizes	34
3.3	Three consecutive line sizes	43
3.4	Two non-consecutive line sizes	57
3.5	On the existence of linear spaces with certain line sizes	65
3.6	Exercises	66

3.7	Research problems	67
3.8	References	67
4	**Semiaffine linear spaces**	**68**
4.1	I-Semiaffine linear spaces	68
4.2	$\{0,1\}$-Semiaffine linear spaces	69
4.3	$\{0,s,t\}$-Semiaffine linear spaces with non-constant point degree	72
4.4	$\{s-1,s\}$-Semiaffine linear spaces with constant point degree	80
4.5	$\{0,1,s\}$-Semiaffine linear spaces with constant point degree	84
4.6	$\{0,1,2\}$-Semiaffine linear spaces	88
4.7	Exercises	94
4.8	Research problems	95
4.9	References	95
5	**Semiaffine linear spaces with large order**	**96**
5.1	Introduction	96
5.2	Embedding linear spaces with constant point degree	104
5.3	Embedding linear spaces with arbitrary point degree	106
5.4	Exercises	118
5.5	Research problems	118
5.6	References	118
6	**Linear spaces with few lines**	**119**
6.1	Restricted linear spaces	119
6.2	The combinatorial approach	125
6.3	The algebraic approach	131
6.4	Exercises	134
6.5	Research problems	135
6.6	References	135
7	***d*-Dimensional linear spaces**	**136**
7.1	The definition	136
7.2	Problems in d-dimensions	137
7.3	π-Spaces	138
7.4	Local structures	146
7.5	'Global' conditions	153
7.6	A final embedding theorem	160
7.7	Exercises	164

7.8 Research problems 165
7.9 References 165

8 Group action on linear spaces 167
8.1 Reasons for this chapter and some background 167
8.2 Point and line orbits 168
8.3 Relationships between transitivities 172
8.4 Characterization theorems 176
8.5 Exercises 187
8.6 Research problems 187
8.7 References 187

Appendix 189

Notation index 212

Subject index 213

In memory of
PAUL de WITTE
1931–1980

Preface

The aim of this monograph is to give a comprehensive and up-to-date presentation of the theory of finite linear spaces. For the most part, we take a combinatorial approach to the subject, but in the final chapter group theory is introduced.

The text is designed as a research resource for those working in the area of finite linear spaces, while the structure of the book also encourages its use as a graduate level text. At the end of each chapter, there is a section of exercises designed to test and extend a student's knowledge of the material in that chapter. There is also a research problem section containing current open problems which can be tackled with the aim of producing a thesis or a journal publication.

In the first chapter, constructions of affine and projective spaces are reviewed, and the fundamental results on finite linear spaces are given. Chapters 2 through 6 cover the work done on the major problem areas in linear spaces taking the 'planar' view: classification of linear spaces with given parameters, embeddability of linear spaces in 'suitably small' projective planes. In Chapter 7 we consider problems of embedding higher dimensional linear spaces in projective spaces. Finally, in Chapter 8, assumptions are introduced on the collineation groups of linear spaces, and the recent results on characterization are presented.

There are several people we wish to thank for their assistance, encouragement and patience while this book was being written. Jean Doyen, Université Libre de Bruxelles, provided us with the appendix of small linear spaces which appears at the end of the book. We also wish to thank him, and the numerous other people whose work is included in our text and reference sections, for the contributions they have made in the area, and for their support for the assimilation of their results in the present monograph. Klaus Metsch, Eindhoven, is responsible not only for much of the research described in Chapter 6, but also for organizing it in the form in which it appears here. Eva Loewen did a great deal of the typing of the original manuscript and patiently made additions and alterations over several years. Xiaomin Bao painstakingly proofread the final version of the manuscript.

Finally, we wish to extend our sincere thanks to David Tranah and to Lauren Cowles, editors at Cambridge University Press, who have endured our changes and deadline extensions with good humour and have maintained a constant supply of support for this project.

1

The essentials

1.1 Definitions and examples

Definition: A *linear space* is a pair $S = (p, \mathcal{L})$ consisting of a set p of elements called *points* and a set $\mathcal{L}$ of distinguished subsets of points, called *lines* satisfying the following axioms:

(L1) Any two distinct points of S belong to exactly one line of S.
(L2) Any line of S has at least two points of S.
(L3) There are three points of S not on a common line.

It is clear that (L3) could be replaced by an axiom (L3)′: There are three lines of S not incident with a common point. In any case, (L3) and (L3)′ are 'non-triviality' conditions. The readers should quickly describe those systems satisfying (L1) and (L2) but not (L3). These are called *trivial linear spaces*.

Points will usually be denoted by the lower case letters $p, q, s, \ldots, x, y, z$, and lines by the upper case letters $L, M, N, \ldots, X, Y, Z$.

The line through the distinct points p and q will be denoted by pq. If two distinct lines L and M intersect in some point, then their (unique) point of intersection will be denoted by $L \cap M$.

We shall use 'geometric' language such as 'a point is on a line', 'a line goes through a point' and so forth, rather than confining ourselves to precise set-theoretic terminology.

Throughout this book we shall be restricting ourselves to *finite linear spaces*, that is, to linear spaces for which the point set is finite. From now on, unless otherwise indicated, *S will always denote a finite linear space*.

We use v and b to denote respectively the number of points and of lines of **S**. For any point p, r_p denotes the number of lines on p. For any line L, k_L denotes the number of points on L. We shall also refer to r_p and k_L as the *degree* or *size* of the respective point p and line L. The terms *i-point* and *i-line* may also be used to refer respectively to a point or a line of degree i.

The numbers v, b, r_p and k_L for all p and L will be called the *parameters* of S.

We proceed to give a number of classes of examples of linear spaces. Some of these we shall draw. For the sake of clarity, we shall never include the 2-point lines in the diagram. There can never be any confusion here, as the lines of size 2 can be reconstructed uniquely.

Complete graphs

A linear space with v points in which any line has just two points is a *complete graph* and is often denoted by K_v. Of course, if $v < 3$, these give trivial linear spaces. In accordance with the above convention, the picture of a complete graph K_v is just a set of v points with no lines (Figure 1.1.1).

Near-pencils

Let $v \geq 3$ be an integer. A *near-pencil* on v points is the linear space having one $(v - 1)$-line and v-1 2-lines. The near-pencil on five points is shown in Figure 1.1.2. Note that $b = v$.

Stars

Let $k_1, k_2, \ldots, k_s$, $s \geq 2$, be integers ordered in such a way that $3 \leq k_1 \leq k_2 \leq \ldots \leq k_s$. A *$(k_1, k_2, \ldots, k_s)$-star* (Figure 1.1.3) is the linear space S described as follows.

Figure 1.1.1. K_7.

Figure 1.1.2. A near-pencil.

Figure 1.1.3. A (3,3,4)-star.

i. There is a particular point p of S of degree s such that the degrees of the lines on p are precisely $k_1, k_2, \ldots, k_s$.
ii. Any line not on p is a 2-line.

A (k_1, k_2)-star is also called a *(k_1, k_2)-cross*.

1.2 Affine spaces and projective spaces

In this section we present possibly the most important classes of linear spaces.

An *affine plane* is a linear space A which satisfies the following axiom.

(A) If the point p is not on the line L, then there is a unique line on p missing L.

The axiom (A) is equivalent to the famous 'parallel postulate' of Euclid. (See *Euclides* 1956, a translation of Euclid's *Elements* by T. L. Heath.) Real Euclidean 2-dimensional space is therefore an example of an affine plane.

We show in the theorem below that any vector space over a skew-field can be used to construct an affine plane.

Theorem 1.2.1. Let F be a skew-field and denote by V a 2-dimensional vector space over F. Define the structure $A = A(V)$ as follows. The **points** *of A are the elements of V. The* **lines** *of A are the (right) cosets of 1-dimensional subspaces of V, that is, the sets $U + v$ where U is a 1-dimensional subspace. Then A is an affine plane.*

PROOF. First we show that A is a linear space.

(L1) Let v and w be distinct points of A. Then v and w are both contained in the coset $\langle v - w\rangle + w$, where $\langle v - w\rangle$ denotes the subspace spanned by the vector $v - w$. Now let $U + x$ be any line containing v and w. Then $v, w \in U + x$ imply $U + v = U + w = U + x$. Hence $v - w \in U$ and so $U = \langle v - w \rangle$. Thus $\langle v - w \rangle$ is the *unique* line of A on v and w.

(L2) Any 1-dimensional subspace has at least two elements, and so any line has at least two points.

(A) Fix a line $U + v$ and a point $w \notin U + v$. Let $U' + v'$ be a line on w which does not meet $U + v$ and with $U' \neq U$. Then V is the sum of U and U'. We may therefore let $v = u_1 + u_1'$ and $v' = u_2 + u_2'$. But now

$$u_2 + u_1' = u_2 + v - u_1 = u_2 - u_1 + v \in U + v$$

and

$$u_2 + u_1' = v' - u_2' + u_1' = u_1' - u_2' + v' \in U' + v',$$

contradicting our assumption that $U + v$ and $U' + v'$ miss each other.

Thus $U = U'$ is the only possibility. Now $w \in U' + v' = U + v'$ implies $U + v' = U + w$, and therefore $U + w$ is the unique line on w missing $U + v$. □

Choosing F in Theorem 1.2.1 to be $GF(2)$, $GF(3)$ and R respectively yields the two smallest examples of affine planes and also the real Euclidean plane. Figure 1.2.1 below shows the affine plane on nine points.

The definition of affine plane has no finitary restrictions. Indeed, as

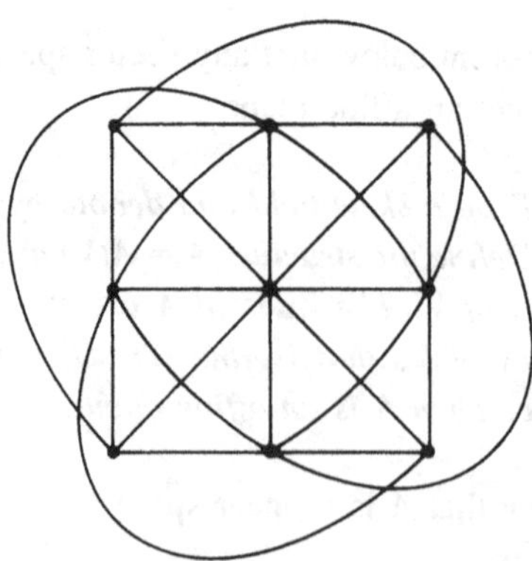

Figure 1.2.1. The affine plane of order 3.

we have just seen, many examples of infinite affine planes do exist. However, for us in this text, the major interest is in the finite case, and therefore we now supply a list of equivalent characterizations of affine planes on a finite number of points.

Proposition 1.2.2. Let S be a linear space on a finite number v of points. Then S is an affine plane if and only if there is an integer $n \geq 2$ *such that* $r_p = n + 1$ *and* $k_L = n$ *for all points p and lines L.*

PROOF. The proof is an easy exercise. □

Proposition 1.2.3. Let S be a linear space on a finite number v of points. Then S is an affine plane if and only if there is a positive integer $n \geq 2$ *such that* $v = n^2$ *and* $k_L = n$ *for all lines L.*

PROOF. Let L and H be lines of S and suppose p is a point not in $L \cup H$. Using axiom (A), p is on $|L| + 1 = |H| + 1$ lines, and so $|L| = |H| = n$, say. If there is no such point p, it is easily seen that L and H are disjoint and that all lines have two points in this case.

Now let L and H be distinct intersecting lines. Using (A), each point not on L is on a unique line missing L but meeting H. Hence $v = n^2$.

In the other direction, for any point p not on the line L, p is easily seen to be on precisely $n + 1$ lines, and hence (A) is satisfied. □

We leave the proof of the next proposition as an exercise.

Proposition 1.2.4. Let S be a linear space on a finite number v of points. Then S is an affine plane if and only if there is a positive integer n such that $b = n^2 + n$ *and* $k_L = n$ *for every line L.*

It is clear from the preceding three propositions that in any finite affine plane A there is an integer n such that $v = n^2$, $b = n^2 + n$, $r_p = n + 1$ and $k_L = n$ for all points p and lines L. This number will be called the *order* of A.

In any affine plane one sees immediately that each line uniquely determines a class of lines having the property that each point is on a unique line of the class. This can also happen in other linear spaces, and so we introduce the following definition.

Let S be any linear space. A (partial) *parallel class* in S is a set of lines of S with the property that each point of S is on (at most) a unique element of the set. Two lines in the same parallel class are said to be *parallel*, and we write $L \parallel M$.

A *parallelism* of S is a set of parallel classes of S such that every line of S is contained in a unique element of the set.

We now use this notion of a 'parallelism' in order to generalize the idea of an affine *plane* to that of an affine *space*. Although we have not mentioned it explicitly, an affine plane of order $n > 2$ is a plane in the sense that it 'is generated by' three non-collinear points. This is false for the affine plane of order $n = 2$. (In fact, 2-lines can be rather unpleasant at any time!) For more details, see A. Beutelspacher (1983) and L. M. Batten (1986). The following definition(s) of affine space is due to H. Lenz (1954).

An affine space of order $n \geq 3$ is a linear space A such that the following conditions hold.

(A1) A has a parallelism.

(A2) Let L and L' be distinct lines of a common parallel class and p a point of neither line. Let M and M' be lines on p such that M meets both L and L', and M' meets L. Then M' must also meet L'.

(A3) $n \geq 3$ is the maximum number of points per line.

It can then be shown that each line has precisely n points. We leave this as an exercise. It is easy to see that affine planes of order greater than 2 are also affine spaces. We should, for completeness, also define an affine space of order 2.

An *affine space of order 2* is a linear space A such that (A1) above holds and also the following.

(A2)′ Let p, q and s be non-collinear points. Then the unique line on s of the parallel class on pq and the unique line on q of the parallel class on ps have a common point.

(A3)′ Every line has precisely two points.

It is probably not clear at present why we wish to distinguish so carefully between these two definitions of affine space. Why should not linear spaces satisfying (A1), (A2)′ and (A3)′ be classified as 'affine spaces'? The answer lies in the theory connecting affine spaces to projective spaces, which we shall see below. Also there is a small group of interesting linear spaces which satisfy (A1), (A2) (which becomes vacuous in view of (A3)′) and (A3)′ but which do not satisfy (A2)′.

A *projective plane* is a linear space P satisfying the following axioms.

(P1) Any two distinct lines have a point in common.
(P2) There are four points, no three of which are on a common line.

A non-trivial space satisfying (P1) but not (P2) is called a *degenerate projective plane*. Clearly, any near-pencil is a degenerate projective plane; conversely, every degenerate projective plane is a near-pencil.

As was the case with affine planes, we can construct many projective planes by using a vector space over a skew-field.

Theorem 1.2.5. Let F be a skew-field and denote by V a 3-dimensional vector space over F. Define the structure $P = P(V)$ as follows. The **points** *of P are the 1-dimensional subspaces of V. The* **lines** *of P are the 2-dimensional subspaces of V. The point p 'is in' the line L if the corresponding 1-dimensional subspace lies in the corresponding 2-dimensional subspace. Then P is a projective plane.*

PROOF. We show first that P is a linear space. Let $\langle X\rangle$ denote the subspace spanned by the set of vectors X. If X is a singleton, we take the liberty of omitting the set brackets.

(L1) If $\langle v\rangle$ and $\langle w\rangle$ are distinct 1-dimensional subspaces of V, then v and w are linearly independent. So $\langle v\rangle$ and $\langle w\rangle$ are contained in $\langle\{v, w\}\rangle$, a 2-dimensional subspace. Conversely, any 2-dimensional subspace on $\langle v\rangle$ and $\langle w\rangle$ must contain $\langle\{u, w\}\rangle$. It follows that $\langle\{v, w\}\rangle$ is the unique line through $\langle v\rangle$ and $\langle w\rangle$.

(L2) Axiom (L2) is clear since a 2-dimensional subspace is spanned by two vectors.

(P2) and (L3) Axiom (L3) follows from (P2). There are three linearly independent vectors u_1, u_2 and u_3 of V. Then the vectors u_1, u_2, u_3, $u_1 + u_2 + u_3$ are linearly dependent. It follows that no three of the four points $\langle u_1\rangle$, $\langle u_2\rangle$, $\langle u_3\rangle$, $\langle u_1 + u_2 + u_3\rangle$ are on a common line.

Note that up to now we have only used $\dim V \geq 3$.

(P1) Let U and U' be distinct 2-dimensional subspaces of V. Then

$\langle U \cup U' \rangle$ is a subspace of V with dimension greater than 2. Hence $\dim\langle U \cup U' \rangle = 3$. So $\dim(U \cap U') = \dim U + \dim U' - \dim \langle U \cap U' \rangle = 2 + 2 - 3 = 1$. Thus the lines U and U' intersect in a point of P. □

The smallest projective plane has seven points. (The fact that it is the smallest and is the unique projective plane on seven points is left as an exercise.) It is called the *Fano Plane* after G. Fano, 1871–1952, and appears in Figure 1.2.2.

Projective planes form a particularly beautiful class of linear spaces because of the dual role played by the points and lines. That is, the points satisfy the same basic axioms as lines, and vice versa. To be more formal, the *dual* of a statement about points and lines is obtained from the statement by interchanging the words 'point' and 'line'. Suitable adjustments to the English are then made so as to make it readable. For example, the dual of (L2) is 'Any point of S is on at least two lines of S'. We leave as an exercise the fact that the duals of (L1), (L2), (L3), (P1) and (P2) hold in any projective plane.

For finite projective planes we now list a number of propositions giving characterizations. The proofs of Propositions 1.2.7 and 1.2.8 are left to the reader.

Proposition 1.2.6. Let S be a linear space on a finite number v of points. Then S is a projective plane if and only if there is an integer $n \geq 2$ such that $r_p = n + 1$ and $k_L = n + 1$ for all points p and lines L.

PROOF. Suppose S is a projective plane on v points, v finite. Let L and M be distinct lines. By axiom (P1) there is a point not on either line. By using axiom (P1) it is now easy, via such a point, to set up a 1–1 cor-

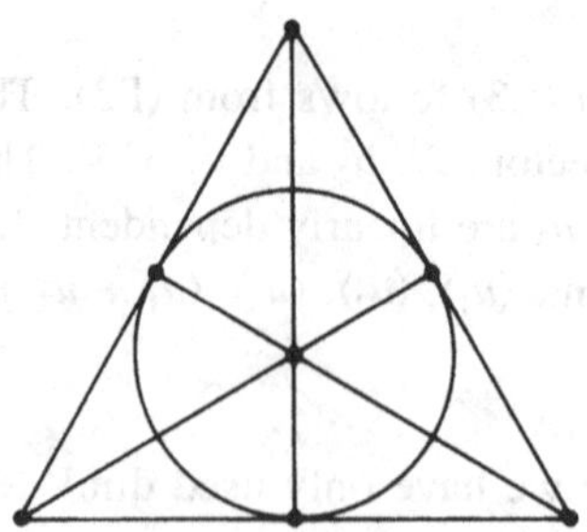

Figure 1.2.2. The Fano plane.

respondence between the points on L and those on M. Hence $k_L = n + 1$ is constant, and using (P1), $n \geq 2$. Applying (P1) again yields $r_p = n + 1$.

Assume that $r_p = n + 1$ and $k_L = n + 1$, $n \geq 2$, for all points p and lines L. Then axioms (P1) and (P2) follow immediately. □

Proposition 1.2.7. Let S be a linear space on a finite number v of points. Then S is a projective plane if and only if there is an integer $n \geq 2$ such that $v = n^2 + n + 1$ and $k_L = n + 1$ for all lines L.

Proposition 1.2.8. Let S be a linear space on a finite number v of points. Then S is a projective plane if and only if there is an integer $n \geq 2$ such that $b = n^2 + n + 1$ and $k_L = n + 1$ for every line L.

It follows from the above propositions that in any finite projective plane P, $v = b = n^2 + n + 1$ and $r_p = k_L = n + 1$ for all points p and lines L, and for some integer $n \geq 2$. We call n the *order* of P.

To give now a definition of *projective space* we need the following ideas.

Let S be any linear space. A subset X of the points of S with induced lines is a (linear) *subspace* of S if for any two distinct points p and q of X, the line pq is a subset of X. It is easy to see that X becomes a (possibly trivial) linear space in its own right.

Again, let S be an arbitrary linear space and let X be a subset of the points of S. The *linear subspace generated by* X is the intersection of all linear subspaces of S containing X. (Since S itself contains X, such a subspace always exists.) We write $\langle X\rangle$ for the linear subspace generated by X.

A *basis* of the linear subspace S' of S is a set of points X such that $\langle X\rangle = S'$ and such that $\langle X - \{x\}\rangle$ is a proper subset of S' for every $x \in X$.

The *dimension* of a linear subspace S' of S is $\max\{|X|-1 \mid X \text{ a basis of } S'\}$.

If the dimension of S' is d we say S' is *d-dimensional*.

A *plane* is a 2-dimensional linear space.

A *hyperplane* of S is a maximal proper linear subspace of S.

A *projective space* is a linear space such that every plane is a projective plane. We say that a linear space is *generalized projective* if each plane is either a projective plane or a degenerate projective plane.

Proposition 1.2.9. All lines in a projective space have the same number of points.

In view of Proposition 1.2.9, it is possible to define the *order* of a projective space to be one less than the number of points per line, so as to coincide with our definition of order for projective planes.

It is a fundamental, but non-trivial, result that every projective or affine space which properly contains a projective, respectively affine, plane can be constructed from a d-dimensional vector space over a skew-field for some d, in a way which easily generalizes our construction for $d = 2$. However, this is not true for $d = 2$ itself. For more details, we refer the reader to Batten (1986), Beutelspacher and Rosenbaum (1992) or D. R. Hughes and F. C. Piper (1973).

Projective and affine spaces constructed over skew-fields will be denoted by $PG(d, n)$ and $AG(d, n)$ respectively, where d denotes the dimension and n the order of the space. Such spaces are called Desarguesian.

1.3 The connection between affine and projective spaces

There is a very intimate relationship between affine and projective spaces:

Theorem 1.3.1. Any affine space of order n (possibly infinite) can be extended to a projective space of order n. Conversely, any projective space of order n contains an affine space of order n.

We shall not give the proof in full detail, as it is lengthy. For a detailed version, see Beutelspacher (1983). To show that an affine space can be extended to a projective one, each parallel class is made to correspond to a 'new point' or 'point at infinity', and new lines are introduced. The new structure consisting of all points of A and the new points, and of the new lines, is proved to be a projective space. Each line of A gets one additional point: the point at infinity corresponding to the unique parallel class of which it is a member (A1). The set of new points alone forms a hyperplane of the projective space.

In the other direction, removing any hyperplane from a projective space of order n is seen to leave us with an affine space of order n.

It would be natural now to suppose, because of this relationship between affine and projective spaces, and because of the definition of projective space, that any plane of an affine space is an affine plane. For the order 2 case, the problem lies in the fact that the affine plane of order 2 is not a plane according to our definition! (However, affine planes of order not 2 really *are* planes.) Axiom (A2)′ for the order 2 case was designed to overcome this problem. However, the order 2 case is not the only strange case. In 1960 M. Hall Jr. gave an example of a linear space in which all planes are affine of order 3, but which is itself not an affine space. This led to an infinite class of such examples. In the meantime, F. Buekenhout (1969) proved that if every plane of a linear space is affine of order at least 4, then the space is indeed affine.

Theorem 1.3.1 is an example of an embedding theorem.

In general, a linear space $S = (p, \mathcal{L})$ is said to be *embedded in* a linear space $S' = (p', \mathcal{L}')$ if $p \subseteq p'$ and $\mathcal{L} \subseteq \mathcal{L}'$. S is *embeddable* in S' if some linear space S^* embedded in S' is isomorphic to S.

Throughout this book we shall be proving embedding theorems where the space S' is a projective plane or, less often, a projective space of higher dimension. The particular case of Theorem 1.3.1 for an affine plane A gives us some insight into how to go about finding such an embedding for an arbitrary linear space. In this case the parallelism determines an equivalence relation on the set of lines such that L and L' are related precisely when $L = L'$ or $L \cap L' = \phi$. We obtain a projective plane as in the theorem, by adjoining a line to A.

1.4 The history of finite linear spaces

The notion of 'linear space' as it is defined in Section 1.1 is due to P. Libois, who first introduced it in a discursive paper on the subject, published in the *Bulletin de la Société Mathématique de Belgique* in 1964. In fact, only our axiom (L1) is required there, but (L2) is also mentioned. Libois attempted in this first article on linear spaces to describe the essential role 'linearity' has played throughout history in what were then considered to be the four 'fundamental' linear spaces ('*les quatre espaces linéaires primitifs*'): Euclidean space, Archimedean time, the platonic system of natural numbers, and the weights and measures of Greek mathematics.

What was probably the first genuine result for finite linear spaces, now generally referred to as the de Bruijn–Erdös–Hanani Theorem, appeared much earlier than Libois's paper. For the interesting history of this theorem, we refer the reader to P. de Witte (1975). Here, we shall limit ourselves to a few comments. The theorem states that in any finite linear space it is always the case that $b \geq v$; moreover, $b = v$ if and only if S is a projective plane or a near-pencil. According to T. Motzkin (1951), the inequality was first conjectured by himself 'not later than 1933' and was first proved by H. Hanani in 1938. However, Hanani's first published proof seems to have appeared in 1951. In the meantime, N. G. de Bruijn and P. Erdös published a proof of the full theorem in 1948, this probably being the first proof to appear in print.

The de Bruijn-Erdös-Hanani theorem, which we shall also refer to as the Fundamental Theorem (for finite linear spaces), and which we shall prove in the next section, gives a *characterization result*. The characterization problem consists in deriving structural properties from numerical ones, in particular, from information about the parameters of the linear space. Thus for $b = v$, we know the structure of S. It is natural then to ask about the situation $b = v + 1$, $b = v + 2$ and so on. The general case $b \leq v + \sqrt{v}$ was handled by J. Totten (1976a), and a shorter proof of Totten's result was given by J. C. Fowler (1984). More recently, K. Metsch (1986) has generalized the earlier papers, and we present the results with proofs in Chapter 6.

Characterization results are often of the type: 'S can be embedded in a projective plane of a certain order and is therefore the complement of a certain configuration in this plane'. Thus we might be led to ask what is perhaps a 'deeper' question about linear spaces: that is, is S embeddable in a finite projective plane (of a particular order)? M. Hall Jr. (1943) gave a construction embedding almost any linear space in a projective plane; but the resultant projective plane is generally infinite. On the other hand, in the preceding section we saw how to embed affine planes in projective planes *of the same order*. We are much more interested in finding constructive embeddings similar to this latter type. They will, however, depend on a 'natural parallelism' in the space, giving rise to an equivalence relation. The linear spaces equipped with such a parallelism are precisely those satisfying the following semiaffinity condition.

(SA) For any point p and line L, there is at most one line on p missing L.

In 1962 N. Kuiper and P. Dembowski proved (see Dembowski 1968) that any finite linear space satisfying (SA) is embeddable in a projective plane the least order of which (apart from near-pencils) is $[\sqrt{v}]$ (the integral part of $\sqrt{v}$). Thus the question arises: if we are to embed S in P, can we anticipate the smallest order possible for P? The semiaffine result suggests $[\sqrt{v}]$, which coincides with the result of the previous section for the embedding of an affine plane in a projective one. Reinforcing this, Totten (1974, 1976) has used the same order, describing it as the unique positive integer n such that $n^2 \leq v < (n + 1)^2$. This we therefore shall call the *square order* of S. On the other hand, Dembowski (1968) defines the *(Dembowski) order of a linear space* as one less than the largest value of r_p, and uses this number as the yardstick for embeddability. Using the first definition, P. de Witte (1976) and D. R. Stinson (1982) obtained an impressive embedding result.

Very recently, Beutelspacher and Metsch (1986, 1987) has proved the following fundamental result. He shows that any finite linear space in which the (Dembowski) order n is larger than a particular function of the smallest line size, is embeddable in a projective plane of order n. We present the details in Section 5.3. This result complements the slightly earlier result of L. Teirlinck (1982, 1986), that a d-dimensional linear space S, $d \geq 3$, in which all lines are 'sufficiently large' is embeddable in a projective space of dimension d and order that of S.

In the treatment of linear spaces in this text, we shall consider both the characterization and embedding points of view. The major embedding results are very recent, as mentioned above. To keep a historical perspective, we shall begin with the major characterization results (though these themselves will not be presented in 'chronological' order). It is evident that a major step in the proofs of these results is an embedding in a projective plane of appropriate order. One thus becomes suspicious that 'large classes' of linear spaces should in fact be projective spaces minus certain configurations. We emphasize that in most cases, knowing that an embedding exists is not sufficient to characterize the space. But in fact, the very structure of the embedding can help in obtaining a characterization.

Once the notion of 'linear space' was made formal by Libois in Brussels in the early sixties, there followed an 'incubation period' of several years during which a handful of researchers at Université Libre de Bruxelles – among them Buekenhout, R. Deherder, J. Doyen and de Witte – began the development of the theory. In the 1970s and 1980s people in other parts of the world began to recognize the fundamental importance of the concept, not only as part of geometry, but also because of deep applications which were found in coding theory, combinatorics and design theory. Capri, Italy, saw the first international meeting devoted to the theory of finite linear spaces in May 1991. Thus at the present time, large numbers of people in many parts of the globe are working on linear spaces. The chapters of the present text are designed to outline the different approaches this research has taken. In Chapters 1 through 6 the 0-

and 1-dimensional objects (points and lines) play the leading role. But in the final two chapters, higher-dimensional objects also appear on the stage.

1.5 Some basic results and the Fundamental Theorem

In this section we introduce some of the basic counting arguments and results for finite linear spaces which will be encountered again and again throughout the book.

Proposition 1.5.1. Let S be a linear space, p a point, L a line with $p \notin L$. *Then* $r_p \geq k_L$.

PROOF. This follows immediately from (L1). □

Proposition 1.5.2. In any finite linear space

$$\sum_{L \in \mathcal{L}} k_L = \sum_{p \in \mathcal{P}} r_p.$$

PROOF. The proof consists in counting point-line indidences in two different ways. The sum on the left counts the number of points on each line, line by line. The sum on the right counts the number of lines per point, point by point. Clearly, these are the same. □

The 'counting in two ways' technique of Proposition 1.5.2 will be met often.

Proposition 1.5.3. In any finite linear space

$$\sum_{L \in \mathcal{L}} k_L (k_L - 1) = v(v - 1).$$

PROOF. We count the number of ordered pairs of points in two different ways. First, counting line by line, since any two distinct points uniquely determine a line (L1), we get the left-hand side. Then, clearly, the number of ordered pairs of points is only the right-hand side. □

Proposition 1.5.4. Let S be a finite linear space. Then

$$\sum_{p \in \mathcal{P}} r_p (r_p - 1) \leq b(b - 1)$$

with equality if and only if any two lines meet.

PROOF. The left-hand side counts the number of intersecting line pairs, whereas the right-hand side is the number of all line pairs.

We now come to the proof of the Fundamental Theorem of finite linear spaces – or rather two of the proofs. As has already been mentioned, there are several proofs in the literature. Here, we give the first proof to appear – that due to de Bruijn and Erdös (1948), and also, for comparison's sake, the shortest known proof, which is due to J. Conway. (See J. G. Basterfield and L. M. Kelly 1968.)

Theorem 1.5.5. (Fundamental Theorem). Let S be a finite non-trivial linear space. Then $b \geq v$. Moreover, equality holds if and only if S is a projective plane or a near-pencil.

PROOF OF DE BRUIJN AND ERDÖS. Suppose $b \leq v$. We shall show that $v = b$ and that S is a projective plane or a near-pencil. We already know that for near-pencils and projective planes, we have $v = b$.

Let L_1 and L_2 be respectively the longest and second longest lines (where $k_{L_1} = k_{L_2}$ is possible). For notational purposes, we write $k_i = |L_i|$, $i = 1,2$. For all points $p \notin L_1$ we then have $r_p \geq k_1$ by Proposition 1.5.1. For all points p of L_1, except possibly $L_1 \cap L_2$, we have $r_p \geq k_2$ by the same proposition. And if $p = L_1 \cap L_2$ exists, then $r_p \geq k_b$ where k_b is the length of the shortest line. Hence

$$\sum_{p \in p} r_p \geq (v - k_1)k_1 + (k_1 - 1)k_2 + k_b.$$

On the other hand,

$$\sum_{L \in \mathcal{L}} k_L \leq k_1 + (b - 2)k_2 + k_b$$

$$\leq k_1 + (v - 2)k_2 + k_b$$

since $b \leq v$. Then, using Proposition 1.5.2, we get

$$(v - k_1)k_1 + (k_1 - 1)k_2 + k_b \leq \sum_{p \in p} r_p = \sum_{L \in \mathcal{L}} k_L \leq k_1 + (v - 2)k_2 + k_b,$$

which implies

$$v(k_1 - k_2) \leq (k_1 + 1)(k_1 - k_2).$$

If $k_1 \neq k_2$, then $v \leq k_1 + 1$ implies that S is a near-pencil. So suppose $k_1 = k_2$. Therefore,

$$(v - 1)k_1 + k_b \leq \sum_{p \in p} r_p = \sum_{L \in \mathcal{L}} k_L \leq (v - 1)k_1 + k_b,$$

which in consequence becomes a system of equalities. In particular, this implies that $k_1 = r_p$ for all r_p except possibly for the smallest value of r_p, which we shall call r_v. But we note that $r_v = k_b$.

In addition, we obtain $v = b$ and all $k_L = k_1$ except possibly k_b.

By Proposition 1.2.6 it suffices to show that $k_1 = r_v$.

But let q be such that $r_q = r_v$. If S is not a near-pencil, there are at least two lines not on q. Hence $r_v = r_q \geq k_1 = r_p$ for all $p \neq q$. Since $r_p \geq r_q$ by definition, we get $k_1 = r_v$. □

CONWAY'S PROOF. We first note that $b - r_p > 0$ and $b - k_L > 0$ for all points p and lines L.

Suppose $b \leq v$. It follows that $(v - k_L)/(b - k_L) \geq v/b$ for any line L. We introduce the following notation.

For any point p and line L,

$$\delta(p,L) = \begin{cases} 0 & \text{if } p \notin L \\ 1 & \text{if } p \in L. \end{cases}$$

Note that

$$\sum_{p \in \mathcal{P}} \delta(p,L) = k_L$$

for fixed L and

$$\sum_{L \in \mathcal{L}} \delta(p,L) = r_p$$

for fixed p.

Now,

$$v = \sum_{p \in \mathcal{P}} \frac{b - r_p}{b - r_p} = \sum_{p \in \mathcal{P}} \sum_{L \in \mathcal{L}} \frac{1 - \delta(p,L)}{b - r_p}.$$

Then

$$\frac{1 - \delta(p,L)}{b - r_p} \geq \frac{1 - \delta(p,L)}{b - k_L}, \tag{1}$$

since this is certainly true for $p \in L$, and for $p \notin L$ we have $r_p \geq k_L$ by Proposition 1.5.1.

So

$$v \geq \sum_{p \in p} \sum_{L \in \mathcal{L}} \frac{1 - \delta(p,L)}{b - k_L} = \sum_{L \in \mathcal{L}} \sum_{p \in p} \frac{1 - \delta(p,L)}{b - k_L}$$

$$= \sum_{L \in \mathcal{L}} \frac{v - k_L}{b - k_L} \geq \sum_{L \in \mathcal{L}} \frac{v}{b} = v.$$

Thus we have equality everywhere, implying $(v - k_L)/(b - k_L) = v/b$, and so $v = b$. In addition, for $p \notin L$, equality in equation (1) implies $r_p = k_L$, and so any two lines meet. It follows that S is a projective plane or a near-pencil. □

1.6 Designs

A finite linear space S in which each line is incident with exactly k points is called a $2 - (v,k,1)$ *design*. (Clearly this notation was conceived for more general structures. For the definition and theory of $t - (v,k,\lambda)$ designs, see Hall 1967.) If S has constant line size, it is easy to see that it must have constant point size also. It is also easy to compute v as a function of r and k, and in addition, using the principle of 'double counting', to obtain the equation $vr = bk$. We summarize these results in the next proposition but leave detailed proofs to the reader.

Proposition 1.6.1. Let S be a $2 - (v, k, 1)$ design. Then each point is on a constant r lines where $v = r(k - 1) + 1$ and $vr = bk$.

The conditions of Proposition 1.6.1 are of course necessary for the existence of a design, but they are not always sufficient. However, for $k = 3$, in 1847 T. Kirkman proved that they are indeed sufficient to ensure the existence of a design. Then in 1965 and 1972, H. Hanani was able to prove this for $k = 4$ and $k = 5$.

The case $k = 6$ is less straightforward. For instance, it is not yet known whether or not there is a $2 - (46, 6, 1)$ design. However, it *is* known that there is no $2 - (36, 6, 1)$ design.

But in 1972, a spectacular result was published by R. Wilson. His theorem says that for any k there exists an integer v_0 such that for all $v \geq v_0$ a $2 - (v,k,1)$ design exists if and only if the above necessary conditions are satisfied. We say that the necessary conditions are *asymptotically sufficient*.

Even more interesting for the theory of linear spaces is the following result, also due to Wilson (1975): Given integers $k_1, \ldots, k_s \geq 2$, the equation of Proposition 1.5.2 for some not necessarily distinct integers $r_1, \ldots, r_v$, such that for all i there is a choice $k_{i_1}, k_{i_2}, \ldots, k_{i_{r_i}}$, of the set $\{k_1, k_2, \ldots, k_s\}$ satisfying

$$v - 1 = \sum_{j=1}^{r_i} (k_{i_j} - 1),$$

is asymptotically sufficient for the existence of a linear space with v points and line degrees $k_1, k_2, \ldots, k_s$.

The designs $2 - (v,3,1)$ were much studied by J. Steiner, 1796–1863, and we shall refer to them as *Steiner triple systems*. The notation $S(2,3,v)$ is also used in this case.

Note that affine and projective planes are $2 - (n^2, n, 1)$ and $2 - (n^2 + n + 1, n + 1, 1)$ designs respectively. We have already seen how to construct examples for n a prime power (since skew-fields always have prime power order). There are many examples of affine and projective planes constructed from weaker algebraic structures, but all such planes also have prime power order. It is not known whether there can exist a projective or affine plane of non-prime power order. The smallest non-prime power is 6, and it is known that no projective or affine plane of this order exists. The next case is 10, and this was eliminated by computer in 1988. (See C. Lam, L. Thiel and S. Swiercz 1989.)

From our point of view, designs are 'known' objects in the sense that when classifying linear spaces if we can show we have a design, then we are finished.

1.7 Another example of a linear space

In this section we consider what J. Totten (1976a,b) and L. M. Batten and J. Totten (1980) have called the *Nwankpa–Shrikhande plane*. This is a linear space on 12 points and 19 lines with constant point size 5, each point being on one 4-line and four 3-lines. Thus, this linear space has the same parameters as the affine plane of order 4 less one line and all its points. J. Totten (1976b) proved that there are precisely two linear spaces with these parameters. Totten also proved that though the former space is not embeddable in a projective plane of order 4, it *is* embeddable in a projective plane of order 5.

Thus, knowing the parameters of a linear space is not enough to determine its structure completely. (Nor is it possible to embed *every* linear space of Dembowski order k in a projective plane of order k.)

There are four non-isomorphic projective planes of order 9 (see G. Kolesova 1989). On the other hand, the projective planes of orders 2, 3, 4, 5, 7 and 8 can be shown to be unique.

The *extended Nwankpa–Shrikhande plane* is the Nwankpa–Shrikhande plane with one additional point on all 4-lines. The *Nwankpa plane* is the unique linear space on 11 points having one 5-point line and fifteen 3-point lines, which is obtained from a K_6 by adding five points corresponding to the partitions of the K_6 into three 2-point lines.

One parting word about notation. If q is a subset of p, then by $S - q$ we mean the linear space with points, the points of S not in q, and with lines the restrictions of lines of S to points not in q as long as such lines have at least two points.

If q consists of a single point $q = \{q\}$, we often write $S - q$ instead of $S - \{q\}$, and we say that S is *punctured*.

1.8 Exercises

1. Explain the impact of (L2) in the definition of linear space.
2. Show that the $v \times b$ incidence matrix of a finite linear space has rank v.
3. Compute the parameters of a $(k_1, k_2, \ldots, k_s)$-star.
4. Prove Proposition 1.2.2.
5. Prove Proposition 1.2.4.
6. Show that an affine plane of order not 2 has dimension 2.
7. Show that all lines in any affine space have the same number of points.
8. Give a characterization of all degenerate projective planes.
9. Show that the Fano plane is the smallest projective plane, and the only one on seven points.
10. Prove that the dual statements of (L1), (L2), (L3), (P1) and (P2) hold in any projective plane.
11. Prove Propositions 1.2.7 and 1.2.8.
12. Prove Proposition 1.2.9.
13. Let P be any projective space. Show that removing a hyperplane from P results in an affine space.
14. Prove Proposition 1.6.1.

15. Prove that any two planes of a projective (affine) space are isomorphic. (See Batten 1986 for a definition of 'isomorphic'.)

1.9 Research problems

1. For which pairs (v, b) does there exist a linear space with v points and b lines? (See Theorem 8.2.4.)
2. For given b, find a function determining the number of linear spaces with b lines.

1.10 References

Basterfield, J. G. and Kelly, L. M. (1968), A characterization of sets of n points which determine n hyperplanes. *Proc. Cambr. Phil. Soc. 64*, 585–588.

Batten, L. M. (1986), *Combinatorics of Finite Geometries*. Cambridge University Press, Cambridge, New York, Melbourne.

Batten, L. M. and Totten, J. (1980), On a class of linear spaces with two consecutive line degrees, *Ars Comb. 10*, 107–114.

Beth, T., Jungnickel, D. and Lenz, H. (1985), *Design Theory*. B.I.-Wissenschaftsverlag, Mannheim, Wien, Zürich.

Beutelspacher, A. and Metsch, K. (1986), Embedding finite linear spaces in projective planes, *Ann. Discrete Math. 30*, 39–56.

Beutelspacher, A. and Metsch, K. (1987), Embedding finite linear spaces in projective planes II, *Discrete Math. 66*, 219–230.

Beutelspacher, A. (1983), *Einführung in die endliche Geometrie* I, II. Bibliographisches Institut, Mannheim, Vienna, Zürich.

de Bruijn, N. G. and Erdös, P. (1948), On a combinatorial problem. *Indag. Math. 10*, 421–423, and *Nederl. Akad. Wetensch. Proc. Sect. Sci. 51*, 1277–1279.

Buekenhout, F. (1969), Une caractérisation des espaces affins basée sur la notion de droite. *Math. Z. 111*, 367–371.

Dembowski, P. (1962), Semiaffine Ebenen. *Arch. Math. 13*, 120–131.

Dembowski, P. (1968), *Finite Geometries*. Springer-Verlag, New York.

Fowler, J. C. (1984), A short proof of Totten's classification of restricted linear spaces. *Geom. Ded. 15*, 413–422.

Hall Jr., M. (1943), Projective planes. *Trans. Am. Math. Soc. 54*, 229–277, and correction *65* (1949), 473–474.

Hall Jr., M. (1960), Automorphisms of Steiner triple systems. *IBM J. Res. Develop. 4*, 460–472.

Hall Jr., M. (1967, revised 1986). *Combinatorial Theory*. Blaisdell, Waltham, Mass.

Hanani, H. (1951), On the number of straight lines determined by n points. *Riveon Lematematika 5*, 10–11 (in Hebrew).

Hanani, H. (1965), A balanced incomplete block design. *Ann. Math. Statist. 36*, 711.

Hanani, H. (1972), On balanced incomplete block designs with blocks having five elements. *J.C.T.* (A) *12*, 184–201.

Heath, Sir Thomas L. (1956), *Euclides*, 2nd ed. (translation), Dover, New York.

Hughes, D. R. and Piper, F. C. (1973), *Projective planes*. Springer-Verlag, New York, Heidelberg, Berlin.

Hughes, D. R. and Piper, F. C. (1985), *Design Theory*. Cambridge University Press, Cambridge, New York, Melbourne.

Kirkman, T. (1847), On a problem in combinations. *Cambr. and Dublin Math. J. 2*, 191–204.

Kolesova, G. (1989), A complete search for projective planes of order 9. Master's Thesis, Concordia University, Montreal.

Lam, C. W. H., Thiel, L. and Swiercz, S. (1989), The non-existence of finite projective planes of order 10. *Can. J. Math. 41*, 1117–1123.

Lenz, H. (1954), Zur Begründung der analytischen Geometrie. *Bayerische Akad. Wiss. 2*, 17–22.

Libois, P. (1964), Quelques espaces linéaires. *Bull. Soc. Math. Belgique 16*, 13–32.

Metsch, K. (1986), *Einbettung linearer Räume in projektive Ebenen*. Ph.D. thesis, Johannes-Gutenberg-Universität, Mainz.

Motzkin, T. (1951), The lines and planes connecting the points of a finite set. *Trans. Amer. Math. Soc. 70*, 451–464.

Stinson, D. R. (1982), A short proof of a theorem of de Witte. *Ars Comb. 14*, 79–86.

Teirlinck, L. (1982), Embeddability properties of linear spaces and planar spaces into projective spaces. *Bull. Soc. Math. Belgique 34*, 69–78.

Teirlinck, L. (1986), Combinatorial properties of planar spaces and embeddability. *J. Comb. T.* (A) *43*, 291–302.

Totten, J. (1974), *Classification of restricted linear spaces*. Ph.D. thesis, University of Waterloo.

Totten, J. (1976a), Classification of restricted linear spaces. *Can. J. Math. 28*, 321–333.

Totten, J. (1976b), Embedding the complement of two lines in a finite projective plane. *J. Austral. Math. Soc.* (A) *22*, 27–34.

Wilson, R. (1972a), An existence theory for pairwise balanced designs I. Composition theorems and morphisms. *J. Comb. Theory* (A) *13*, 220–245.

Wilson, R. (1972b), An existence theory for pairwise balanced designs II. The structure of PBD-closed sets and the existence conjectures. *J. Comb. Theory* (A) *13*, 246–273.

Wilson, R. (1975), An existence theory for pairwise balanced designs III. Proof of the conjectures. *J. Comb. Theory* (A) *18*, 71–79.

de Witte, P. (1975), Combinatorial properties of finite linear spaces, II. *Bull. Soc. Math. Belgique 27*, 115–155.

de Witte, P. (1976), On the embeddability of linear spaces in projective planes of order *n* (unpublished manuscript).

2
Complementation

2.1 Aim of the chapter

The aim of this chapter is to show explicitly how certain linear spaces can be embedded in a projective plane. Among such structures are the complements of two lines, of a triangle, of a hyperoval and of a Baer subplane. Here, the notion of a pseudo-complement is crucial. Suppose that we remove a set X of a projective plane P of order n. Then we obtain a linear space P-X having certain parameters (i.e., the number of points, the number of lines, the point- and line-degrees). We call any linear space which has the same parameters as P-X a *pseudo-complement* of X in P.

We have already encountered the notion of a pseudo-complement, namely the *pseudo-complement of one line*. This is a linear space with n^2 points, $n^2 + n$ lines in which any point has degree $n + 1$ and any line has degree n. We know that this is an affine plane, which is a structure embeddable into a projective plane of order n. Another example may help to clarify the above definition. A *pseudo-complement of two lines* in a projective plane of order n is a linear space having $n^2 - n$ points, $n^2 + n - 1$ lines in which any point has degree $n + 1$ and any line has degree $n - 1$ or n. (More precisely, the n-lines form a parallel class.) We shall show that for $n > 4$ any such structure may be obtained from a projective plane of order n by removing two lines. So, the leitmotif of this chapter is: pseudo-complements (at least the most natural ones!) are complements!

We shall consider finite incidence structures S consisting of points and lines (which are non-empty subsets of the point set) such that through any two distinct points there is exactly one line. Loosely speaking, such a structure is a linear space in which lines of degree 1 are allowed. Consequently, we shall use all the already developed terminology of linear spaces.

2.2 The details

A pair $(p, \mathcal{L})$ which satisfies (L1) and additionally the condition any point is on exactly r lines is called an *(r, 1)-design*.

From now on we suppose that S is a linear space or an $(r, 1)$-design, as the context demands.

The underlying idea of the construction of an embedding is the following. We "extend" our linear space S by adjoining new points in order to reach eventually a projective plane.

Definition: Let S be an $(r, 1)$-design and denote by p a point of S. The incidence structure $S' = S - p$ consists of the points of S which are distinct from p and of those lines of S which are incident with a point $\neq p$. Then S' is again an $(r, 1)$-design. We call S an *extension* of S'.

An $(r, 1)$-design S^* is said to be *s times extendible to S* if S^* is obtained from S by removing one at a time, s points in the way described above. (In the terminology of Chapter 1, S^* is embedded in S.)

Proposition 2.2.1. An $(n + 1, 1)$-design with $b = n^2 + n + 1$ lines and s mutually intersecting lines of degree n is s times extendible.

PROOF. For every n-line N we define

$$\Pi_N = \{N\} \cup \{X | X \text{ a line disjoint to } N\}.$$

Since each point has degree $n + 1 = k_N + 1$, each point outside N lies on exactly one line which is parallel to N. This shows that Π_N is a parallel class. Because N meets n^2 other lines, $|\Pi_N| = n + 1$.

Suppose that N and G are two different n-lines which meet. Then G meets n lines of Π_N, so $|\Pi_G \cap \Pi_N| = 1$.

Since there are s mutually intersecting n-lines, there are parallel classes $\Pi_1, \ldots, \Pi_s$ any two of which have a unique line in common. We let the lines of Π_j intersect in a new point p_j, $j = 1, \ldots, s$. Since Π_j is a parallel class, p_j is connected to every old point by a unique line.

Furthermore, the unique common line in Π_j and Π_k is the line through the two new points p_j and p_k. This shows that we have obtained a linear space. Since every parallel class Π_j has $n + 1$ lines, this linear space is an $(n + 1, 1)$-design. □

Proposition 2.2.2. Suppose that S is an $(n + 1, 1)$-design with $b = n^2 + n + 1$ lines and $v = n^2 + n + 1 - s$ points. Then the number b_n of *n-lines is at least $s(n + 2 - s)$ with equality if and only if every line has degree $n - 1$, n, or $n + 1$.*

PROOF. We have, as in Propositions 1.5.2 and 1.5.3,

$$v(n + 1) = \sum_{p\in \mathcal{P}} r_p = \sum_{L\in \mathcal{L}} k_L \text{ and } v(v - 1) = \sum_{L\in \mathcal{L}} k_L(k_L - 1).$$

It follows that

$$\sum_{L\in \mathcal{L}} (n + 1 - k_L)(n - 1 - k_L) = s(s - 2 - n).$$

Since

$$-b_n \leq \sum_{L\in \mathcal{L}} (n + 1 - k_L)(n - 1 - k_L),$$

we have

$$s(n + 2 - s) \leq b_n.$$

Equality holds if and only if

$$\sum_{L\in \mathcal{L}} (n + 1 - k_L)(n - 1 - k_L) = b_n,$$

that is if and only if there are only lines of degree $n + 1$, $n - 1$ and n. □

Now we can already embed a class of linear spaces in projective planes.

Proposition 2.2.3. (Vanstone 1973; de Witte 1975). Suppose that S is a linear space and every point has degree at most $n + 1$. If $v \geq n^2$, $n \geq 2$, then S can be extended to a projective plane of order n.

PROOF. Each point is on n or $n + 1$ lines, and each line on n or $n + 1$ points. Each $(n + 1)$-point is only on n-lines. Any $(n + 1)$-line is on all n-points. It follows that any n-point p corresponds to a parallel class of n, n-lines of $S - p$. By adding a new point p^* corresponding to this parallel class, and a new line pp^*, both points p and p^* become $(n + 1)$-points. We may therefore assume that S is an $(n + 1, 1)$-design.

Clearly $k_L \leqq n + 1$ for all lines L, with equality if and only if L intersects every other line. If there is a line of degree $n + 1$, then $b = 1 + k_L \cdot n = n^2 + n + 1$, since S is an $(n + 1, 1)$-design.

If $v = n^2 + n + 1$, then S is a projective plane (cf. Proposition 1.2.7). On the other hand if $v < n^2 + n + 1$, then by Proposition 2.2.2 there exists an n-line. In view of Proposition 2.2.1, S can be extended. Repeating this procedure completes the proof.

If every line has degree at most n, then in view of $v \geqq n^2$ and $r_p = n + 1$ for all points p, it follows that every point lies only on lines of degree n. Hence, S is an affine plane of order n (see Proposition 1.2.2). □

Corollary 2.2.4. Suppose that S is an $(n + 1, 1)$-design with $n^2 + n + 1$ lines which has s mutually intersecting n-lines. If $v + s \geqq n^2$, then S can be embedded in a projective plane of order n.

PROOF. Propositions 2.2.1 and 2.2.3. □

2.3 The pseudo-complement theorems

We have seen that parallel classes play a crucial role in extending linear spaces. So far, they have been induced by n-lines. If there are no n-lines, it is much harder to prove the existence of a parallel class. For example, in order to show that a line L of degree $n - 1$ in an $(n + 1, 1)$-design lies in a parallel class, one has to show that the set of lines parallel to L can be partitioned into two sets of mutually parallel lines. For this, the following proposition will be useful.

Proposition 2.3.1. Let S be an $(n + 1, 1)$-design and denote its number of lines by $b = n^2 + n + 1 - z$. Suppose that there exists a line L of degree $n - 1$ which is parallel to lines L_1, L_2 and L_3. If $n + 1 - d_j$ is the degree of L_j, $j = 1, 2, 3$, and if any two of the lines L_j meet, then

$$n \leqq d_1 \cdot d_2 + d_1 \cdot d_3 + d_2 \cdot d_3 - d_1 - d_2 - d_3 - z.$$

PROOF. Let M be the set of lines missing L. Because there are $k_L \cdot n = n^2 - n$ other lines, we have $|M| = b - 1 - (n^2 - n) = 2n - z$.

Let M_j be the set of lines of $M - \{L_j\}$ which meet L_j. Since every point of L_j lies on two lines parallel to L, we have $|M_j| = n + 1 - d_j$.

For $j \neq k$, every line of $M - (M_j \cup M_k)$ is parallel to L_j and L_k.

Next we claim that the number of lines parallel to L_j and L_k equals $d_j \cdot d_k - z$. In fact, this number is the number $b - 2$ of all lines not equal to L_j, L_k, minus the number of lines intersecting L_j or L_k plus the number of lines intersecting L_j and L_k. Thus we get

$$b - 2 - k_{L_j} \cdot n - k_{L_k} \cdot n + n - 1 + (n - d_j)(n - d_k) = d_j \cdot d_k - z.$$

Since L misses L_j and L_k, it follows that $|M| - |M_j \cup M_k| \leqq d_j \cdot d_k - z - 1$. Hence

$$|M_j \cap M_k| = |M_j| + |M_k| - |M_j \cup M_k|$$
$$\leq |M_j| + |M_k| - |M| + d_j \cdot d_k - z - 1 = d_j \cdot d_k - d_j - d_k + 1.$$

This implies:

$$|M_1 \cup M_2 \cup M_3| \geqq |M_1| + |M_2| + |M_3| - |M_1 \cap M_2|$$
$$- |M_1 \cap M_3| - |M_2 \cap M_3|$$
$$\geqq 3n - d_1 \cdot d_2 - d_1 \cdot d_3 - d_2 \cdot d_3 + d_1 + d_2 + d_3.$$

In view of $M_j \subseteq M$ and $|M| = 2n - z$, the assertion follows from

$$3n - d_1 \cdot d_2 - d_1 \cdot d_3 - d_2 \cdot d_3 + d_1 + d_2 + d_3 \leqq |M| = 2n - z. \qquad \square$$

Theorem 2.3.2. (Totten 1976; cf. also Oehler 1975 and Batten 1991.) Every pseudo-complement S of two lines in a projective plane of order n ≧ 5 is in fact a complement.

PROOF. By definition S has $n - 1$ lines of degree n and n^2 lines of degree $n - 1$, and every point lies on a unique line of degree n and on n lines of degree $n - 1$. In particular, the n-lines are mutually parallel and every n-line meets every line of degree $n - 1$.

Let L be any line of degree $n - 1$. Denote by H a line parallel to L, and by M the set of lines which meet H and miss L. It follows that the lines of M are mutually parallel. (Assume that there were two intersecting lines H_2, H_3 in M. Then H, H_2 and H_3 intersect each other mutually and we may apply Proposition 2.3.1. Since $d_1 = d_2 = d_3 = 2$ and $z = 2$, we get

$$n \leqq 3 \cdot 2 \cdot 2 - 3 \cdot 2 - 2 = 4.)$$

Consequently, $\Pi = M \cup \{L\}$ is a set of n mutually parallel lines of degree $n - 1$. Since the number of points is $v = n(n - 1)$, Π is a parallel class. Since every point outside L is contained in two lines parallel to L, the set Π' consisting of L and the lines parallel to L which are not in Π is also a parallel class. We have $|\Pi'| = n$ and $\Pi \cap \Pi' = \{L\}$.

Now we let the lines of Π (or Π') intersect in a new point ∞ (or ∞', respectively). Furthermore, we adjoin the two new lines $\{\infty\}$ and $\{\infty'\}$ of degree 1. The resulting structure is an $(n + 1, 1)$-design D with $v + 2 = n^2 - n + 2$ points and $b + 2 = n^2 + n + 1$ lines. Since the point ∞

lies on n lines of degree n of D, Proposition 2.2.1 shows that D can be embedded in an $(n + 1, 1)$-design D' having $n^2 + 2$ points.

Then Proposition 2.2.3 implies that D' is embeddable in a projective plane P of order n. It follows that S is embedded in P. Obviously, S is the complement of two lines. □

Corollary 2.3.3. If S is a pseudo-complement of two lines in a projective plane of order $n \geq 3$, then S is a projective plane of order n less two lines and all their points, or S is the Nwankpa–Shrikhande plane.

In the corollary, the Nwankpa–Shrikhande plane arises from the case $n = 4$. For the details, see J. Totten (1976).

Results similar to that of Theorem 2.3.2 exist for the cases where S is the complement of α lines all on a common point. The results can be found in R. C. Mullin and S. A. Vanstone (1976), R. C. Mullin, N. M. Singhi and S. A. Vanstone (1977) and P. de Witte (1983).

A *triangle* in a linear space is a set of three lines, not all on a common point, along with all points of those lines.

The following theorem (in particular its proof) is the prototype of many embedding theorems. So, study the proof carefully.

(Ralston's original proof used $n \geq 25$.)

Theorem 2.3.4. (Ralston 1981.) Every pseudo-complement of a triangle in a projective plane of order $n > 6$ is in fact a complement.

PROOF. Let S be the pseudo-complement of a triangle. In other words, S is an $(n + 1, 1)$-design with $v = (n - 1)^2$ points and $b = n^2 + n - 2$ lines such that every line has degree $n - 2$ or $n - 1$. We call the lines of degree $n - 1$ *long* and the lines of degree $n - 2$ *short*. It follows that

(1) every point lies on 3 long lines and on $n - 2$ short ones, which implies that
(2) there are $3(n - 1)$ long lines and $(n - 1)^2$ short ones. Moreover,
(3) a long line is parallel to $n - 2$ long lines and to $n - 1$ short lines, and
(4) there is a unique line which misses two intersecting long lines.

Fix a long line L. We claim that the long lines which are parallel to L are mutually parallel. Assume to the contrary that L is parallel to the intersecting long lines L_1 and L_2. Then, by (4), every line parallel to L meets L_1 or L_2. Since L_1 (and L_2, respectively) meets $n - 1$ lines parallel to L and since L is parallel to only $2n - 3$ lines, it follows that there is a (unique) line L_3 which is parallel to L and which meets L_1 and L_2. Because L_3 has degree $n - 1$ or $n - 2$, Proposition 2.3.1 implies $n \leqq 6$, a contradiction.

Hence, the long lines which are parallel to L are mutually parallel. Together with L they form a set Π of $n - 1$ mutually parallel long lines. In view of $v = (n - 1)^2$, Π is a parallel class.

Let L' be an arbitrary line of Π. In view of $|\Pi| = n - 1$, every line which is parallel to L' and which is not contained in Π has degree $n - 2$ and meets every line of $\Pi - \{L'\}$. Every point outside L' lies on two lines parallel to L', one of which is contained in Π. It follows that every point outside L' has a unique short line parallel to L'. In other words, L' together with the short lines parallel to L' form a parallel class. Furthermore, every line other than L' of this parallel class meets every line of $\Pi - \{L'\}$.

Consequently, there are $n - 1$ mutually disjoint parallel classes $\Pi_1, \ldots, \Pi_{n-1}$ each consisting of a long line of Π and the $n - 1$ short lines parallel to it. Since every short line misses a unique line of Π, every short line is contained in exactly one of the parallel classes Π_j.

We let the lines of Π_j intersect in a new point ∞_j and we adjoin the new line $\{\infty_1, \ldots, \infty_{n-1}\}$ to S. Since the Π_j are mutually disjoint, the structure S' obtained in this way is again a linear space. We claim that S' is the pseudo-complement of two lines in a projective plane of order n. To see this, note that each short line of S has degree $n - 1$ in S' since it is contained in one of the sets Π_j. Each line of Π has degree n in S' and every other long line of S still has degree $n - 1$ in S'. Also the new line has degree $n - 1$. This shows that S' is the pseudo-complement of two lines in a projective plane of order n.

Corollary 2.3.3 shows that S' is in fact a complement. By our construction, S can therefore be embedded in a projective plane of order n and is the complement of a triangle. □

A *k-arc* in a projective plane of order n is a set of k points no three of which are collinear. (A k-arc can be thought of as a complete graph embedded in the plane.) It is not difficult to show that $k \leqq n + 1$ if n is odd and $k \leqq n + 2$ if n is even (see, for instance, J. W. P. Hirschfeld 1979). An $(n + 2)$-arc in a plane of (necessarily) even order n is also called a *hyperoval*. A hyperoval has the property that any line intersects it in 0 or 2 points. So, a *pseudo-complement of a hyperoval in a projective plane of order* n is a linear space with $v = n^2 - 1$ points, $b = n^2 + n + 1$

lines, constant point-degree $n + 1$ and lines of degree $n + 1$ and $n - 1$. A last application of Proposition 2.3.1 is

Theorem 2.3.5. Every pseudo-complement of a hyperoval in a projective plane of order $n > 6$ can be embedded into a projective plane as the complement of a hyperoval.

PROOF. Let S be the pseudo-complement of a hyperoval in a projective plane of order n. We call the lines of degree $n + 1$ *long* and the other lines *short*. Every point is contained in $(n + 2)/2$ short lines and in $n/2$ long lines. The number of short lines is $\frac{1}{2} \cdot (n + 2)(n + 1)$ and the number of long lines is $\frac{1}{2}n(n - 1)$. Furthermore, a short line is parallel to $2n$ other short lines and a long line meets every other line.

Fix a short line H and denote by M the set of the $2n$ lines parallel to H. It follows from Proposition 2.3.1 that if M were to contain a triangle, then $n \leqq 6$.

Let L_1 and L_2 be intersecting lines of M; denote by M_1 the set of lines of M which meet L_2 and by M_2 the set of lines of M which meet L_1. Since M contains no triangle, M_1 and M_2 consist of mutually parallel lines. We have $|M_j| = n - 1$ and $L_j \in M_j$. Furthermore, $M_1 \cap M_2 = \phi$ (again because M contains no triangle).

Let G_1 and G_2 be the two lines of $M - (M_1 \cup M_2)$. We claim that each line of M_1 is parallel to at most two lines of M_2. (Each line of M meets $n - 1$ other lines of M. Then every line of M_1 meets at least $n - 3$ lines of M_2. Note that $M = M_1 \cup M_2 \cup \{G_1, G_2\}$ and M_1 consist of mutually parallel lines.)

The line G_1 meets $n - 1$ other lines of M. One of these lines may be G_2 but at least $n - 2$ of them are in $M_1 \cup M_2$. Therefore, without loss of generality, G_1 meets at least $\frac{1}{2}(n - 2) > 2$ lines of M_2. Hence, if L is an arbitrary line of M_1, then L meets a line of M_2, which also meets G_1. Since M has no triangles, this implies that G_1 is parallel to L. So, G_1 is parallel to every line of M_1.

Consequently, $\Pi := M_1 \cup \{H, G_1\}$ is a set of mutually parallel lines with $|\Pi| = n + 1$. In view of $v = n^2 - 1 = |\Pi| \cdot (n - 1)$, Π is a parallel class. Therefore we obtain again an $(n + 1,1)$-design if we let the lines of Π intersect in a new point. By Proposition 2.2.3, this design and therefore also S can be embedded in a projective plane of order n. □

In fact, this result was originally proved by R. C. Bose and S. S.

Shrikhande (1973), and then generalized greatly, allowing $n \geqq 2$ by P. de Witte (1977).

We want to complete this section with another embedding result. This time we shall again use n-lines to construct parallel classes. Let P be a projective plane of order n. A *subplane* is a projective plane B whose points are points of P, whose lines are lines of P and whose incidence is induced by the incidence of P. Let B be a subplane of P, and denote by m its order. A theorem of R. H. Bruck (see D. R. Hughes and F. C. Piper 1973) says that if $m < n$, then $m \leqq \sqrt{n}$. A subplane of order $\sqrt{n}$ is called a *Baer subplane* of P.

Theorem 2.3.6. Every pseudo-complement S of a Baer subplane can be embedded in a projective plane as the complement of a Baer subplane.

PROOF. By definition, S has $v = n^2 - \sqrt{n}$ points, $b = n^2 + n + 1$ lines, constant point degree $n + 1$ and lines of degree n (*long* lines) and $n - \sqrt{n}$ (*short* lines), for some $n \geq 2$.

Since $r_p = n + 1$ for any point p, any long line L is contained in a unique parallel class $\Pi(L)$. Since L meets n^2 other lines, we have $|\Pi(L)| = b - n^2 = n + 1$.

It can easily be shown that any point is on exactly one short line. Hence the total number of short lines is $b_s = v/(n - \sqrt{n}) = n + \sqrt{n} + 1$. Since there is a total of $b - b_s = n^2 - n$ long lines and since every parallel class of type $\Pi(L)$ has $n - \sqrt{n}$ long lines, it follows that there are exactly $(n^2 - \sqrt{n})/(n - \sqrt{n}) = n + \sqrt{n} + 1$ such parallel classes.

Now we adjoin the parallel classes as new points to S. Since any two parallel classes intersect in exactly one line, we obtain a linear space S'. This linear space has $n^2 - \sqrt{n} + n + \sqrt{n} + 1 = n^2 + n + 1$ points and $n^2 + n + 1$ lines; therefore S' is a projective plane of order n.

In particular, any short line has now $n + 1$ points, so it has $\sqrt{n} + 1$ new points. Similarly, any long line is incident with just one new point. Since there are exactly $n + \sqrt{n} + 1$ new points, the new points together with the short lines form a projective plane of order $\sqrt{n}$.

Thus, Theorem 2.3.6 is proved completely. □

H. Swart and K. Vedder (1981) show that the union of d mutually disjoint subplanes of order m in a finite projective plane of order n is determined by its complement if $d^2 - d < n - m$. A more difficult problem is to characterize such a complement. L. M. Batten and S. S. Sane (1985) gave such a characterization for $d = 2$:

Theorem 2.3.7. Let S be a linear space with the following properties:

1. $v = m^4 - m^2 - 2m - 1$, $12 < m < \infty$.
2. *Each point is on* $m^2 + 1$ *lines.*
3. $b = m^4 + m^2 + 1$.
4. *Each line is on* $m^2 - 1$ *or* $m^2 - m - 1$ *points.*
5. *For any pair L, L′ of disjoint lines on* $m^2 - 1$ *points, and for any point* $p \notin L \cup L'$, *there exists a line on p missing L and L′, but not more than one such line is on* $m^2 - 1$ *points.*

Then S is uniquely embeddable in a projective plane P of order m^2 *and is the complement of the union of two disjoint Baer subplanes of P.*

This result is not true for $m = 8$, as was shown by M. de Resmini (1985). We leave the other values of $m \leq 12$ to the reader as a research problem.

For the sake of completeness, we mention that the complement of a unital (see the exercises for the definition) has been considered by L. J. Dickey (1971) and Beutelspacher (1984) (see also P. J. Cameron 1976), and the complement of a quadrilateral by M. S. Montekhab (1986).

2.4 Exercises

1. Prove the assertions (1), (2), (3), (4) in the proof of Theorem 2.3.4.
2. Consider the case $n < 5$ in Theorem 2.3.2. (See Corollary 2.3.3.)
3. Consider Theorem 2.3.4 for $n \leq 4$.
4. A *unital* is a linear space with $v = m^3 + 1$ and constant line size $k = m + 1$, $m \geq 2$.
   (a) Compute the parameters of a unital.
   (b) A unital embedded in a projective plane **P** of order n is a unital which is a substructure of **P** and has $n\sqrt{n}$ points. Compute the parameters of a unital embedded in a projective plane of order n, and the parameters of its complement.
   (c) Formulate and prove a theorem analogous to Theorem 2.3.5 for pseudo-complements of unitals. (If you get stuck, refer to L. J. Dickey 1971 or A. Beutelspacher 1984.)

2.5 Research problems

1. Analyze the cases $m \leq 12$ of Theorem 2.3.7.

2.6 References

Batten, L. M. (1991). A dual approach to embedding the complement of two lines in a finite projective plane. *J. Austral. Math. Soc.* (A) *51*, 426–435.

Batten, L. M. and Sane, S. (1985). A characterization of the complement of the union of two disjoint Baer subplanes. *Arch. Math. 44*, 569–576.

Beutelspacher, A. (1984). Embedding the complement of a Baer subplane or a unital in a finite projective plane. *Mitt Math. Sem. Univ. Giessen 163*, 189–202.

Bose, R. C. and Shrikhande, S. S. (1973). Embedding the complement of an oval in a projective plane of even order. *Discrete Math. 6*, 305–312.

Cameron, P. J. (1976). *Parallelisms of Complete Designs*. LMS Lecture Notes *23*.

Dickey, L. J. (1971). Embedding the complement of a unital in a projective plane. *Atti del Con. di Geom. Comb. e sue Apple*. Perugia, Sept. 1970.

Hirschfeld, J. W. P. (1979). *Projective Geometries over Finite Fields*. Clarendon Press, Oxford.

Hughes, D. R. and Piper, F. C. (1973). *Projective Planes*. Springer-Verlag, New York.

Montekhab, M. S. (1986). Embedding of finite pseudo-complements of quadrilaterals. *J. Stat. Planning and In. 13*, 103–110.

Mullin, R. C., Singhi, N. M. and Vanstone, S. A. (1977). Embedding the affine complement of three intersecting lines in a finite projective plane. *J. Austral. Math. Soc.* (A) *24*, 458–464.

Mullin, R. C. and Vanstone, S. A. (1976). A generalization of a theorem of Totten. *J. Austral. Math. Soc.* (A) *22*, 494–500.

Oehler, M. (1975). Endliche biaffine Inzidenzebenen. *Geom. Ded. 4*, 419–436.

Ralston, T. (1981). On the embeddability of the complement of a complete triangle in a finite projective plane. *Ars Combinatoria 11*, 271–274.

de Resmini, M. J. (1985). On 2-blocking sets in projective planes. *Ars Combinatoria 20*, 59–69.

Swart, H. and Vedder, K. (1981). Subplane replacement in projective planes. *Equationes Math. 22*, 27–34.

Totten, J. (1976). Embedding the complement of two lines in a finite projective plane. *J. Austral. Math. Soc. 22*, 27–34.

Vanstone, S. A. (1973). The extendibility of (r,l)-designs. Proc. Third Southeastern Conf. on Num. Math., *Utilitas* Math., 409–418.

de Witte, P. (1975). On the embeddability of linear spaces in projective planes of order *n* (unpublished).

de Witte, P. (1977). The exceptional case of a theorem of Bose and Shrikhande. *J. Austral. Math. Soc.* (A) 24, 64–78.

de Witte, P. (1983). Variations on a theorem of Kuiper and Dembowski. *Simon Stevin 57*, 47–59.

3
Line sizes

3.1 Introduction

One of the most natural strictly numerical questions to ask is what can be said if all the line degrees of a linear space S are known. Clearly, this problem will have a reasonable answer only if the set of allowable line degrees is quite small. If there is only one line degree, then S is a design, and in a sense, S is 'known'. We therefore turn to the case of two line sizes. Work on two *consecutive* line sizes was the first to appear, and this was done by L. M. Batten, J. Totten and P. de Witte.

If we place an upper bound on v with respect to the line sizes, then we show below that we are able to say something quite precise about S in terms of its structure relative to a projective plane. The case of two line sizes includes the case of one line size. Thus we also give a precise description of designs with v points where v is bounded above. The results of Sections 3.2 and 3.3 are found in de Witte and Batten (1983) and in Batten and Totten (1980).

In Section 3.3 we present the work of Batten (1980) for three consecutive line sizes.

Section 3.4 deals with two *non*-consecutive line degrees; and in Section 3.5 we briefly describe some general theorems covering a broader class of line size problems.

Let b_k and v_r be the number of k-lines, respectively r-points, in S.

Proposition 3.1.1. If each r-point is on a_r k-lines, then $kb_k = \Sigma_r\, v_r a_r$, as r ranges over all possible values in S.

PROOF. The proposition follows from counting in two different ways the incidences of points and k-lines. □

An *a-pencil* of a linear space is a set of a lines all incident with a fixed point.

3.2 Two consecutive line sizes

We suppose throughout this section that S is a non-trivial finite linear space in which each line has n or $n + 1$ points.

Proposition 3.2.1. (i) $v \geq n^2 - n + 1$; (ii) $(v - 1)/n \leq r_p \leq (v - 1)/(n - 1)$ for any point p; (iii) each point p is on $nr_p - v + 1$ n-lines and $(1 - n)r_p + v - 1$ $(n + 1)$-lines.

PROOF. The proofs of (i) and (ii) are easy. For (iii), let a and c be the number of n- and of $(n + 1)$-lines respectively on the point p. Then $a + c = r_p$. Also $a(n - 1) + cn = v - 1$. So $(a + c)n - a = nr_p - a = v - 1$, implying $a = nr_p - v + 1$ and thus $c = (1 - n)r_p + v - 1$. □

Theorem 3.2.2. If S is a finite linear space with $v \leq n^2 + 2n + 3$, $n \geq 3$, in which each line has n points, then S is a projective plane of order $n - 1$ or an affine plane of order n or a 2-(13,3,1), a 2-(15,3,1), a 2-(25,4,1), a 2-(46,6,1) or a 2-(51,6,1).

PROOF. If each line has n points, then clearly $r_p = (v - 1)/(n - 1)$ for all points p. Counting incident point-line pairs yields $n \leq r_p = (v - 1)/(n - 1) \leq n + 3 + (5/(n - 1))$. So $n \leq r_p \leq n + 5$. If $r_p = n$ or $n + 1$, then $v = r_p(n - 1) + 1$ implies respectively that S is a projective plane of order $n - 1$ (Proposition 1.2.7) or an affine plane of order n (Proposition 1.2.3).

If $r_p = n + 2$, then $nb = (n + 2)v$ (Proposition 1.6.1) $= (n + 2)[r_p(n - 1) + 1]$ implies the contradiction $n|2$.

If $r_p = n + 3$, then $nb = (n + 3)v$ implies $n|6$. If $n = 6$, then we have a 2-(46,6,1) whose existence is still unknown (Hall 1986). If $n = 3$ we have a Steiner triple system on 13 points.

If $r_p = n + 4$ or $n + 5$, then $n - 1 \leq 5$, or $n \leq 6$.

The case $n = 3$ gives $r_p = 7$ or 8 and thus a Steiner triple system on 15 or 17 points. By Proposition 1.6.1, the latter of these does not exist.

The case $n = 4$ gives $r_p = 8$ and we get a 2-(25,4,1) design.

The case $n = 5$ yields $r_p = 9$ and $nb = (n + 4)v$ (Proposition 1.6.1) leads to the contradiction $5|333$.

Finally, if $n = 6$, $r_p = 10$, we get $v = 51$ and S is another 'problematic' design 2-(51,6,1) (Hall 1986). □

In this section, we wish to prove the following result.

Theorem 3.2.3. If S is a non-trivial linear space with $v \leq (n + 1)^2$, $n \geq 5$, in which every line has either n or $n + 1$ points, then S is one of:

(i) a projective plane (respectively affine plane) of order $n - 1$ or n (respectively n or $n + 1$);
(ii) a punctured projective plane (respectively affine plane) of order n (respectively $n + 1$);
(iii) an affine plane of order n together with one point at infinity;
(iv) an affine plane of order $n + 1$ less a line and all its points, except possibly one;
(v) a projective plane of order $n + 2$ less a triangle;
(vi) a 2-(46,6,1) design.

Furthermore, if $v < (n + 1)^2$, $n \geq 3$ and $v \neq 15$, then we need only add to this list the following:

(vii) The Nwankpa–Shrikhande plane;
(viii) a 2-(13,3,1) design;
(ix) the unique linear space on 13 points and 20 lines with one 6-point (the rest having degree 5) and six 4-lines.

The proof of this theorem will be given in its entirety in this section, but will be broken up into a number of propositions. Each proposition will deal with a certain range of values of v. The reader will see that this is a natural division as certain ranges of values for v lend themselves to certain kinds of arguments. *We suppose throughout that $n \geq 3$.*

If all lines have n points, we can use Theorem 3.2.2. If all lines have $n + 1$ points, we may again use Theorem 3.2.2. In this case, however, $r_p = (v - 1)/n$ for all p implies $r_p \leq n + 2$. Clearly $r_p \geq n + 1$. So by the theorem, S is a projective plane of order n or an affine plane of order $n + 1$.

So for the remainder of the section suppose that S contains both n- and $(n + 1)$-lines. It follows that $v \geq n^2$.

Proposition 3.2.4. If $v \leq n^2 + n - 1$, then S is an affine plane of order n.

PROOF. Suppose $v = n^2$ or $n^2 + 1$. Then Proposition 3.2.1 gives

$(v-1)/n \le r_p \le (v-1)/(n-1)$ which, for our values of v, implies $r_p \in \{n, n+1\}$ for each p. If $v = n^2$ then any point not on an $(n+1)$-line L has degree $n+1$; moreover, any line through such a point is an n-line. Thus L is the only $(n+1)$-line. Considering a point p on L we find

$$n^2 = v = k_L + (r_p - 1)(n-1) = n + 1 + (r_p - 1)(n-1);$$

hence, $n - 1 | n^2 - n - 1$, a contradiction.

If $v = n^2 + 1$, since any point not on an $(n+1)$-line has degree $n+1$, any two $(n+1)$-lines intersect. But an $(n+1)$-point is on exactly one $(n+1)$-line and so all $(n+1)$-lines pass through the same point q. Since any point is on at least one $(n+1)$-line, the $(n+1)$-lines are precisely the lines through q. So $S - q$ is an affine plane of order n.

Now assume $n^2 + 2 \le v \le n^2 + n - 2$. Write $v = n^2 + n - d$, $2 \le d \le n - 2$. By Proposition 3.2.1, $n + 1 - (d-1)/n \le r_p \le n + 2 - (d-1)/(n-1)$, which implies $r_p = n + 1$ for each point p. Fix an n-line L. Any point not on L is on a unique line missing L, this line being an n-line as all $(n+1)$-lines meet L. No two of these n-lines meet as otherwise we get some point on at least $n+2$ lines. So this set of n-lines partitions the points, leading to the contradiction $n | n^2 + n - d$.

Finally, consider $v = n^2 + n - 1$. Again using Proposition 3.2.1, we obtain $r_p \in \{n+1, n+2\}$ for all p, and each $(n+1)$-point is on precisely two n-lines while each $(n+2)$-point is on precisely $n+2$ n-lines. Now Proposition 3.1.1 yields $nb_n = 2v_{n+1} + (n+2)v_{n+2} = 2v + nv_{n+2}$. Thus $n | 2v$, or equivalently, $n | 2n^2 + 2n - 2$. So $n | 2$, a contradiction. □

Proposition 3.2.5. If $v = n^2 + n$, then S is a punctured projective plane of order n, an affine plane of order $n+1$ less a line and all its points, or the Nwankpa–Shrikhande plane.

PROOF. From Proposition 3.2.1, $r_p \in \{n+1, n+2\}$ for all points p, and each point is on either one n-line and n $(n+1)$-lines, or on $n+1$ n-lines and one $(n+1)$-line. If $r_p = n+1$ for all p, it is clear that the n-lines partition S, which is therefore a punctured projective plane of order n. Otherwise, letting p be an $(n+2)$-point, since any $(n+1)$-point is incident with $n \ge 3$ lines of degree $n+1$, there is an $(n+1)$-line L missing p. Let L' be the $(n+1)$-line on p.

Suppose that L and L' meet in a point q, which must therefore have degree $n+1$. Let M be the n-line on p missing L. Thus each point of M has degree $n+2$.

Consider a point $q' \ne q$ on L. Since p has degree $n+2$, pq' is an n-

line. Assume for the moment that q' is an $(n + 1)$-point. Then $q'p$ is the only n-line through q', and so there exists an $(n + 1)$-line through q' intersecting an $(n + 1)$-line through q in a point p' of M. This contradicts the fact that p' is an $(n + 2)$-point. Hence q' must be an $(n + 2)$-point, and L is the only $(n + 1)$-line on it.

It follows that any $(n + 1)$-line *not* on q would miss all of the n $(n + 1)$-lines distinct from L' on q. Therefore, all $(n + 1)$-lines pass through q. On the other hand, any point different from q on the n-line through q is on an $(n + 1)$-line which evidently cannot pass through q.

This contradiction implies $L \cap L' = \phi$. So any point on L or L' has degree $n + 2$. Therefore any line meeting L or L', but distinct from either of these lines, is an n-line. We deduce now that any point of S is on at least two n-lines and is therefore of degree $n + 2$. Moreover $b = (n + 1)^2 + n$. By Corollary 2.3.3, we conclude that S is an affine plane of order $n + 1$ less a line and all its points, or S is the Nwankpa–Shrikhande plane. □

Proposition 3.2.6. If $v = n^2 + n + 1$, then S is an affine plane of order $n + 1$ less a line and all its points but one, a 2-(13,3,1) design, or the unique linear space on 13 points and 20 lines with one 6-point and six 4-lines.

PROOF. For each point p using Proposition 3.2.1, $r_p = n + 1$ and each point is only on $(n + 1)$-lines; or $r_p = n + 2$ and each point is on two $(n + 1)$-lines and n n-lines; or $r_p = n + 3$, $n = 3$ and each point lies only on 3-lines.

If all points have degree $n + 1$ then S is a projective plane of order n, which is a contradiction. So we may suppose that points of degree $n + 2$ or $n + 3$ exist.

Assume for the moment that there is no point of degree $n + 3$. Assume also that there is an $(n + 1)$-point p and an $(n + 2)$-point q. Let L be the $(n + 1)$-line through q different from pq. Then every point not on pq or L is joined to q by an n-line and is therefore an $(n + 2)$-point. It follows that L contains at most two $(n + 1)$-points and so we must have at least one $(n + 2)$-point, $q' \neq q$, on L. Since L and $q'p$ are the only $(n + 1)$-lines through q', p is the only $(n + 1)$-point on pq. Suppose p' is a second $(n + 1)$-point. Then $p' \in L$ and pp' is an $(n + 1)$-line. Let s be a point of pp' not on L or pq. Let s' be a point not on L such that ss' is an $(n + 1)$-line different from pp'. Then s' has degree $n + 2$ yet lies on three $(n + 1)$-lines, namely ps', $p's'$ and ss'. This contradiction shows that p is the only $(n + 1)$-point in S.

Since $k_p = n + 1$, each $(n + 1)$-line passing through p meets every other $(n + 1)$-line. Then fixing an $(n + 1)$-line L' on p, we see that each point not on L' is on a unique line missing L', which must be an n-line. We thus get a partition of the points of S corresponding to each $(n + 1)$-line on p, and clearly the partitions are pairwise disjoint since $r_p = n + 1$. So we can adjoin $n + 1$ points at infinity, one corresponding to each partition, and an ideal line M containing precisely the ideal points, to get a new linear space S'. S' has $(n + 1)^2 + 1$ points, and all its lines have degree $n + 1$ or $n + 2$. Now $S' - p$ is clearly an affine plane of order $n + 1$ and can be completed in the standard way to a projective plane of order $n + 1$, say S^*. Then we note that S is obtained from S^* by deleting M^* (the line of S^* corresponding to the line M of S) to get an affine plane of order $n + 1$, and then in addition, deleting a line less one point.

Let us now consider the case that all points have degree $n + 2$. Proposition 3.1.1 yields

$$b_{n+1}\,(n + 1) = 2\,(n^2 + n + 1),$$

implying $(n + 1)|2\,(n^2 + n - 1) = 2\,((n + 1)^2 - n)$, forcing $(n + 1)|2n$, which is impossible.

Finally suppose that $r_p = n + 3$ for some point p. Then $n = 3$ and $r_p = 6$. Also, all points have degree 5 or 6. If all points have degree 6, then S is a Steiner triple system on 13 points. So suppose further that there is at least one 5-point. Since each such point lies on two 4-lines and three 3-lines, we may assume the existence of 4-lines.

Assume first that all 4-lines meet one another. Then there are precisely five 4-lines and these lines contain a total of 10 points, all of which have degree 5. It follows that there are three 6-points and sixteen 3-lines. The three 5-points not joined by 4-lines to a given 5-point are themselves already pair-wise joined by 4-lines. Thus, any 3-line joining two 5-points contains a 6-point and the three 6-points are collinear.

Let $a, b, \ldots, j$ be the 5-points and p, q, s the 6-points. Then we already have the following lines: $\{a,b,c,d\}$, $\{a,e,f,g\}$, $\{b,e,h,i\}$, $\{c,f,h,j\}$, $\{d,g,i,j\}$, $\{p,q,s\}$. Without loss of generality we may also assume the existence of the following lines: $\{a,h,p\}$, $\{a,i,q\}$, $\{a,j,s\}$. Now consider ej. Clearly $s \notin ej$. Suppose that $q \in ej$. Then $q \notin ce$ and since $q \notin ci$, we have $q \in cg$. Also, since $q \notin bj$, bg, we have $q \in bf$. Since $s \notin bf$, bj we see that $s \in bg$. Consequently, $s \notin gh$. But $p \notin gh$. Therefore

$q \in gh$. But this contradicts $q \in cg$. Thus $q \notin ej$. So $p \in ej$. Then $p \notin de$, and since $p \notin dh$ we have $p \in df$. Also, since $p \notin bf, bj$, we have $p \in bg$. Consequently $s \notin bq$ and $s \notin bj$, which together imply $s \in bf$. Therefore $s \notin fi$ and also $q \notin fi$. So $p \in fi$, which contradicts $p \in df$.

Therefore, there are two disjoint 4-lines L_1 and L_2. Then there are four 4-lines meeting both L_1 and L_2, one through each point of L_1 and of L_2. Call them L_3, L_4, L_5, L_6. Clearly L_3 must miss one of L_4, L_5 and L_6. Say L_3 misses L_4. Then the union of L_1, L_2, L_3, L_4 contains 12 points all of which have degree 5 and so are on two 4-lines. Thus L_5 and L_6 must meet both L_3 and L_4 and are disjoint. Let $a = L_1 \cap L_3$, $b = L_1 \cap L_4$, $c = L_1 \cap L_5$, $d = L_1 \cap L_6$, $e = L_2 \cap L_3$, $f = L_2 \cap L_4$, $g = L_2 \cap L_5$, $h = L_2 \cap L_6$, $i = L_3 \cap L_5$, $j = L_3 \cap L_6$, $k = L_4 \cap L_5$, $l = L_4 \cap L_6$. Now examine the points not yet joined to a; f is joined to all of these but p. So $\{a, f, p\}$ is a line. From the remaining points g, h, k, l, we must create two 3-lines on a. It is not difficult to see that these lines must be $\{a,g,l\}$ and $\{a,h,k\}$. Similarly, we get the remaining lines: $\{b,e,p\}$, $\{b,g,j\}$, $\{b,h,i\}$, $\{c,h,p\}$, $\{c,e,l\}$, $\{c,f,j\}$, $\{d,g,p\}$, $\{d,e,k\}$, $\{d,f,i\}$, $\{i,l,p\}$, $\{j,k,p\}$. □

Proposition 3.2.7. The case $n^2 + n + 2 \leq v \leq n^2 + 2n - 1$ is impossible.

PROOF. If $n^2 + n + 2 \leq v \leq n^2 + 2n - 3$ we can write $v = n^2 + n + k$, $2 \leq k \leq n - 3$, which, using Proposition 3.2.1, gives

$$n + 1 + \frac{k-1}{n} \leq r_p \leq n + 2 + \frac{k+1}{n-1}.$$

Since $n \geq 3$ it follows that $r_p = n + 2$ for all p.

Applying Proposition 3.1.1 we obtain

$$b_{n+1} = \frac{(k+1)v}{n+1} \text{ and } b_n = \frac{(n+1-k)v}{n}$$

yielding

$$b = b_n + b_{n+1} = n^2 + 3n + 1 + k \cdot \frac{2n+1-k}{n(n+1)}.$$

Fix an $(n + 1)$-line L. Every point is on a unique line missing L. Thus there are at least $(v/(n + 1)) - 1$ or $n - 1 + (k/(n + 1))$ lines missing L. Counting the lines meeting L we find that $b \geq (n + 1)\ (n + 1)$

$+ n + 1 = n^2 + 3n + 2$. It follows that $((2n + 1 - k)/n(n + 1))k \geq 1$, implying $0 \geq (n - k)(n - (k - 1))$, and so $k - 1 \leq n \leq k$, a contradiction.

Now consider $v = n^2 + 2n - 2$. Since $v \geq n^2 + n + 2$, we may assume $n \geq 4$. Then $r_p \in \{n + 2, n + 3\}$ and an $(n + 2)$-point is on three n-lines and $n - 1$ $(n + 1)$-lines, while an $(n + 3)$-point is on $n + 3$ n-lines and no $(n + 1)$-lines.

By Proposition 3.1.1,

$$b_n = 3(n + 2) - \frac{6}{n} + v_{n+3}$$

and so $n = 6$.

Also, $b_{n+1} = (n - 1)/(n + 1)\ v_{n+2}$.

If $v_{n+3} = 0$, 1, 2 or 3 we have $v_{n+2} = v$, $v - 1$, $v - 2$ or $v - 3$ respectively, which in turn leads to $7|230$, $7|225$, $7|220$, $7|215$, and so a contradiction. Thus $v_{n+3} \geq 4$.

Suppose $v_{n+2} \neq 0$ and therefore $b_{n+1} \geq 5$. Since $v_{n+2} = \frac{7}{5} b_{n+1}$ we have $v_{n+2} \geq 7$.

Let L be an n-line. Since $v_{n+2} \geq 7$ there is an $(n + 2)$-point $p \notin L$, and so L cannot have more than three $(n + 3)$-points.

Suppose L has three $(n + 3)$-points. Fix an $(n + 2)$-point q on L and an n-line, $M \neq L$, on q. Let $p \neq q$ be an $(n + 2)$-point of M. Then p lies on at least four n-lines, which is a contradiction. Thus any n-line contains at most two $(n + 3)$-points.

Finally, fix an $(n + 2)$-point p and label its n-lines L_1, L_2, L_3. All $(n + 3)$-points must be on these lines. Therefore $4 \leq v_{n+3} \leq 6$ and $40 \leq v_{n+2} \leq 42$. Applying Proposition 3.1.1, $(n + 1)b_{n+1} = (n - 1)v_{n+2}$ and we get a contradiction unless $v_{n+2} = 42$, so $v_{n+3} = 4$. The distribution of $(n + 3)$-points on L_1, L_2, L_3 is therefore 2-2-0 or 2-1-1. Assume L_1 has two $(n + 3)$-points, and L_2 at least one. Then the number of points on L_3 which are also on the lines formed by the four $(n + 3)$-points is at most four, and in any case, this leaves an $(n + 2)$-point of L_3 say, not on any of these lines. This $(n + 2)$-point is then on at least four n-lines; a contradiction. We conclude that $v_{n+2} = 0$. Thus, there are no $(n + 1)$-lines, a final contradiction.

For the last case, $v = n^2 + 2n - 1$, Proposition 3.1.1 yields $nb_n = 2v_{n+2} + (n + 2)\ v_{n+3} = 2(n^2 + 2n - 1) + nv_{n+3}$, and so $b_n = 2n + 4 - 2/n + v_{n+3}$. Hence $n|2$, a contradiction. □

Proposition 3.2.8. If $v = n^2 + 2n$, $n \geq 4$, then S is a punctured affine plane of order $n + 1$.

PROOF. Proposition 3.2.1 implies $r_p \in \{n+2, n+3\}$ for all p, an $(n+2)$-point is on one n-line and $n+1$ $(n+1)$-lines, an $(n+3)$-point is on $n+1$ n-lines and two $(n+1)$-lines; moreover, by Proposition 3.1.1,

$$b_n = n + 2 + v_{n+3}, \quad b_{n+1} = v_{n+2} + \frac{2v_{n+3}}{n+1}$$

whence

$$b = n^2 + 3n + 2 + \frac{2v_{n+3}}{n+1}.$$

Suppose $v_{n+3} \neq 0$.

Step 1. Suppose all $(n+3)$-points are on an n-line L. Let p be such a point. Let M be a second n-line on p, and let p_1, p_2, p_3 be $(n+2)$-points of M. Let $q \neq p$ be a second $(n+3)$-point of L. Then all lines joining p_i to q must be $(n+1)$-lines, while q cannot be on three such lines. Thus any n-line misses some $(n+3)$-point.

Step 2. Let L be an n-line and p and q $(n+2)$-points on it. Let s be an $(n+3)$-point not on L. The lines ps and qs must be $(n+1)$-lines. Any other line joining s to L must therefore be an n-line, and then the fact that any point of L other than p and q is on at least two n-lines implies it is an $(n+3)$-point. Hence any n-line contains at most two $(n+2)$-points.

Step 3. We show here that two disjoint n-lines L and M cannot both have two $(n+2)$-points. We calculate v_{n+3}. Write $v_{n+3} = 2(n-2) + d$, where d is the number of $(n+3)$-points not on L and M, and L and M each have two $(n+2)$-points p, p' and q, q' respectively. Let s be an $(n+3)$-point on L. The lines qs and $q's$ must be $(n+1)$-lines and so $d \geq 2(n-4)$. On the other hand, if s is a point not on $L \cup M$, and is not $pq \cap p'q'$ or $pq' \cap p'q$, then at least three of the $(n+1)$-lines ps, $p's$, qs, $q's$ are distinct and so s must be an $(n+2)$-point. Therefore, $d \leq 2$. Combining inequalities yields $n = 4$, $d = 2$, $v_{n+3} = 6$ and from the calculation of b_{n+1} above, we see that $5|12$, which is a contradiction.

Step 4. Let L be an n-line containing the two $(n+2)$-points p and q. Let s be an $(n+3)$-point not on L. Then the lines ps, qs are $(n+1)$-lines, forcing the three lines on s missing L to be n-lines. Let M be one of these three n-lines. Now L and M are disjoint n-lines and L has two $(n+2)$-points, so M has at most one $(n+2)$-point by step 3. Therefore

M has at least three $(n+3)$-points, s, t, u. Now the line qs is an $(n+1)$-line; it may meet the lines tp, up, but even so, it contains a point w different from q and s and not on tp, up, since $n \geq 4$. The lines pt, pu, qt, qu must all be distinct $(n+1)$-lines, forcing the distinct lines tw, uw to be n-lines. So w is an $(n+3)$-point. Finally, let w' be an $(n+3)$-point of L. Let L' be an $(n+1)$-line on w'; it may meet the $(n+1)$-lines ps, qs, pw, but nevertheless, since $n \geq 4$, it contains a point $x \neq s$ and not on any of these lines. The lines sx and wx are both n-lines. So x must be an $(n+3)$-point. But x is on three distinct $(n+1)$-lines L', xp and xq; so we have a contradiction. Hence any n-line contains at most one $(n+2)$-point.

Step 5. Each of the $n+1$ n-lines through a fixed $(n+3)$-point is on at most one $(n+2)$-point by step 4. So $v_{n+2} \leq 2n + n + 1$. On the other hand, if the $(n+1)$-line L is on three $(n+2)$-points p, p', p'', there exist unique n-lines on each of these points. Any point not on one of the four lines above is joined to p, p' and p'' by three distinct $(n+1)$-lines and is therefore an $(n+2)$-point. There are at least v-$(4n-2)$ such points. So $v_{n+2} \geq n^2 - 2n + 2$. Combining the two inequalities we obtain $n = 4$, $v_{n+2} = 13$, $v_{n+3} = 11$. This contradicts $(n+1)|2v_{n+3}$. Thus any $(n+1)$-line has at most two $(n+2)$-points.

Step 6. We prove now that of the $n+1$ n-lines through an $(n+3)$-point at most one contains an $(n+2)$-point. Otherwise we have an $(n+3)$-point p on distinct n-lines each containing $(n+2)$-points q and q', say. Let L be an $(n+1)$-line on p. Any point $s \neq p$, $qq' \cap L$, of L is on at least three $(n+1)$-lines, since qs and $q's$ are distinct $(n+1)$-lines. Hence s is an $(n+2)$-point. Since $n \geq 4$, there are at least three such points s, which contradicts step 5.

Step 7. Applying the results of steps 5 and 6 to a fixed $(n+3)$-point, we see that there are at most five $(n+2)$-points. Suppose p, p', p'' are $(n+2)$-points. By steps 4 and 5 they are not all collinear and the lines formed by them are $(n+1)$-lines. Any point not on pp', pp'', $p'p''$ and not on one of the n-lines on each of p, p', p'' is joined to p, p', p'' by three distinct $(n+1)$-lines and is therefore an $(n+2)$-point. There are at most $6n-3$ points on the six lines above. So $v_{n+2} = v - (6n-3) + 3 = n^2 - 4n + 6 \geq 6$, while we have already established $v_{n+2} \leq 5$.

We conclude that there are at most two $(n+2)$-points.

Step 8. Since $n + 1|2v_{n+3}$ it follows that $v_{n+3} \neq n^2 + 2n$ or $n^2 + 2n - 1$. Hence $v_{n+2} \geq 2$. From step 7 we have $v_{n+2} = 2$ (and $n = 5$).

Let p and p' be $(n + 2)$-points. Let $q \in pp' - \{p, p'\}$. So q is an $(n + 3)$-point. Let $L \neq pp'$ be an $(n + 1)$-line on q. The unique n-lines on p and p' meet L in at most two points. Hence there is, since $n \geq 4$, a point $s \neq q$ on L, s not on the n-lines on p and p'. But ps and ps' are $(n + 1)$-lines, forcing s to be a third $(n + 2)$-point and contradicting step 7.

We conclude then that $v_{n+3} = 0$.

It follows that the n-lines partition the points of L. Join them all in a point at infinity. The resulting structure is an affine plane of order $n + 1$. □

Proposition 3.2.9. If $v = (n + 1)^2$, $n \geq 7$, then S is a projective plane of order $n + 2$ less three non-concurrent lines and all their points.

PROOF. By Proposition 3.2.1 and since $n \geq 5$ we see that for each p, either $r_p = n + 2$ where p is only on $(n + 1)$-lines, or $r_p = n + 3$ where p is on three $(n + 1)$-lines and n n-lines. If all points have degree $n + 2$ we obviously have an affine plane of order $n + 1$, a contradiction. Let us assume then that points of degree $n + 3$ exist. Therefore, there is at least one n-line, and such a line contains only $(n + 3)$-points. So for every $(n + 1)$-line there is at least one $(n + 3)$-point not lying on it. It follows easily that each $(n + 1)$-line has at least $n - 2$ points of degree $n + 3$.

Suppose there are at least three $(n + 2)$-points p, q and s. Let L be any $(n + 1)$-line meeting pq at a point different from p, q, s. Since there is an $(n + 3)$-point on pq which is not on L, there can be at most two $(n + 2)$-points on L which are not on pq. Since $n \geq 5$ there is an $(n + 3)$-point t on L such that $t \notin ps$, qs or pq. This implies that t lies on the four $(n + 1)$-lines L, pt, qt and st, a contradiction.

Consequently there are at most two points of degree $n + 2$. Applying Proposition 3.1.1 twice results in

$$2v_{n+3} = b(n + 1) - (n + 1)^2 (n + 2).$$

Therefore $(n + 1) | 2v_{n+3}$. But $(n + 2)^2 - 2 \leq v_{n+3} \leq (n + 1)^2$ forces $v_{n+3} = (n + 1)^2 = v$.

Theorem 2.3.4 and $n > 6$ finally imply that S is a projective plane of order $n + 2$ less three non-concurrent lines and all their points. □

3.3 Three consecutive line sizes

We are now ready to attack the problem of three consecutive line sizes. The main result of this section is summarized in the following theorem.

Theorem 3.3.1. Let S be a non-trivial linear space with $v \le n^2$ and line range $\{n-1, n, n+1\}$. If $n \ge 23$, S is

(i) *an affine plane of order $n-1$ or n;*
(ii) *an affine plane of order n less a point, a line and all its points, or a line and all its point but one;*
(iii) *an affine plane of order $n-1$ with an additional point at infinity;*
(iv) *a projective plane of order $n-2$ or $n-1$, perhaps less a point in the latter case;*
(v) *a projective plane of order $n+1$ less three lines and all their points, or perhaps retaining one of these points, not the point of intersection, if these lines are concurrent;*
(vi) *a projective plane of order n, less a line and all its points but one, and less one, two or n of the points (and therefore also the line in this last case) of a second line on this point, while retaining the point;*
(vii) *a projective plane of order n less an $(n+1)$-arc or an $(n+2)$-arc, the latter only if n is even;*
(viii) *a projective plane of order n less all points save one of each of two lines, with the point of intersection and the lines themselves deleted.*

Furthermore, if $v = n^2$, $n \ge 8$, or $v < n^2 - 1$, $n \ge 4$, we have only the following additional possibilities:

(ix) *$v = 11$ and S is the Nwankpa plane;*
(x) *$v = 12$ and S is the Nwankpa–Shrikhande plane;*
(xi) *$v = 13$ and S is the extended Nwankpa–Shrikhande plane, a Steiner triple system on 13 points, or the unique linear space on 13 points and 20 lines with one 6-point and six 4-lines;*
(xii) *$v = 46$ and S is a 2-(46,6,1) design.*

We give the proof of the theorem only for $n \ge 23$ although many lower values of n are included. The remaining cases are treated in several exercises at the end of the chapter.

As in the last section, the proof is broken into several propositions. We begin with a preliminary result giving the analogue to Proposition 3.2.1.

Proposition 3.3.2. $v \ge n^2 - n - 1$; $(v-1)/n \le r_p \le (v-1)/(n-2)$ for any point p.

Proposition 3.3.3. If S is a non-trivial linear space with $v \le n^2$ points,

$n \geq 4$, and line range $\{n - 1, n + 1\}$, and $v \neq n^2 - 1$, then S is one of the following:

(i) an affine plane of order $n - 1$;
(ii) a projective plane of order n less two lines and all their points except their meet;
(iii) a projective plane of order $n - 2$;
(iv) one of the problematic block designs 2-(46,6,1);
(v) the extended Nwankpa–Shrikhande plane;
(vi) a Steiner triple system on 13 points; or
(vii) the Nwankpa plane.

If $v = n^2 - 1$, $n \geq 23$, then S is

(viii) a projective plane of order $n + 1$ less three concurrent lines and all their points, or
(ix) a projective plane of even order n less an $(n + 2)$-arc.

PROOF. Theorem 3.2.2 allows us to suppose that there are $(n + 1)$-lines.

Using Proposition 3.3.2 we note that not all lines can be $(n + 1)$-lines, and also, $v \geq (n + 1)(n - 2) + 1 = n^2 - n - 1$.

Case 1. $v < n^2 - 3$. In this case $r_p \leq n + 1$ for all p. Since $r_p = n + 1$ for some p, we may suppose that $v = n^2 - n - 1 + 2m$, where m is the number of $(n + 1)$-lines on an $(n + 1)$-point, and $0 \leq 2m < n - 2$. Let L be an $(n + 1)$-line. Then for $p \notin L$, $r_p = n + 1$ where m of these are $(n + 1)$-lines and $n + 1 - m$ are $(n - 1)$-lines. If $m = 0$, then all lines distinct from L are $(n - 1)$-lines and all lines meet L. Let $q \in L$. Thus $(r_q - 1)(n - 2) + n = v - 1 = (n + 1)(n - 2)$, implying $n - 2 | n$ and so $n = 4$. Thus $S - L$ is a 6-arc, $r_p = 4$ for all $p \in L$, and so S is the Nwankpa plane. If $m = 1$, then through any point not on an $(n + 1)$-line there is precisely one $(n + 1)$-line. Hence the set of $(n + 1)$-lines is the set of lines on a fixed point s. Then $S - s$ is a linear space with $n^2 - n$ points, $r_p = n + 1$ for all p, one of these lines an n-line and n of them $(n - 1)$-lines. By Theorem 3.2.3 S is a projective plane of order $n > 4$ less two lines and all their points except their meet.

So we may suppose $m \geq 2$ and $r_p = n + 1$ for all points p. Clearly then, all $(n + 1)$-lines meet any given line. Let L be any $(n + 1)$-line and M any $(n - 1)$-line. The number of $(n + 1)$-lines meeting M is $(n - 1)m = mn - m$. The number of $(n + 1)$-lines meeting L is $(m - 1)(n + 1) + 1 = mn - n + m$. Hence $2m = n \nless n - 2$, and we have a contradiction.

Case 2. $v = n^2 - 3$. Assume for the moment that $n \geq 6$. Using Proposition 3.3.2, we obtain $r_p = n + 2$, all $(n - 1)$-lines; $r_p = n + 1$, $\frac{1}{2}(n - 2)$ $(n + 1)$-lines, $\frac{1}{2}(n + 4)$ $(n - 1)$-lines; or $r_p = n$, $n - 2$ $(n + 1)$-lines, 2 $(n - 1)$-lines. Any n-point p must be on all $(n + 1)$-lines and is therefore unique. Let q be an $(n + 1)$-point. Then q can be on at most one $(n + 1)$-line, forcing $\frac{1}{2}(n - 2) \leq 1$ or $n \leq 4$, and so a contradiction. Hence n-points do not exist.

If all points have degree $n + 1$, counting incidences of $(n + 1)$-lines in two different ways implies

$$n + 1 | \tfrac{1}{2}(n - 2)(n^2 - 3) = \tfrac{1}{2}n^2(n + 1) - \tfrac{3}{2}n(n + 1) + 3,$$

so that $n + 1 | 3$, a contradiction. Hence, let L be an $(n + 1)$-line and p an $(n + 2)$-point not on L. Then p is on an $(n - 1)$-line M missing L, and each point of M is an $(n + 2)$-point.

If there is an $(n + 2)$-point, q not on M, then as above, q is on an $(n - 1)$-line M' missing L. All lines meeting but distinct from M also meet L; these are $(n + 1)(n - 1) = n^2 - 1$ in number. Only $(n - 1)^2$ of these meet M', leaving $n^2 - 1 - (n - 1)^2 = 2n - 2$ lines meeting L but not M'. But $n^2 - 1$ lines meet M' and L so that the number of lines meeting L is at least $n^2 - 1 + 2n - 2 + 1 = n^2 + 2n - 2$. The precise number of lines meeting L is $n(n + 1) + 1 = n^2 + n + 1$, and so

$$n^2 + 2n - 2 \leq n^2 + n + 1 \quad \text{or } n \leq 3.$$

Thus q does not exist, and all points of $S - M$ are $(n + 1)$-points.

Since no $(n + 1)$-line meets M, each $(n + 1)$-point is on at most two $(n + 1)$-lines, forcing $\frac{1}{2}(n - 2) \leq 2$ or $n \leq 6$.

If $n = 6$, $|S - M| = 28$, each point on two 7-lines and five 4-lines. Also all 7-lines meet each other. Label these {1,2,3,4,5,6,7}, {1,8,9,10,11, 12,13}, {2,8,14,15,16,17,18}, {3,9,14,19,20,21,22}, {4,10,15,19,23,24, 25}, {5,11,16,20,23,26,27}, {6,12,17,21,24,26,28}, {7,13,18,22,25,27, 28}.

The points of M correspond to parallel classes, each containing seven 4-point lines. Without loss of generality, we may choose 1, 14 and 23 to be collinear, forcing the line {1,14,23,28}. Choosing 2, 9 and 24 we find that {2,9,24,27} is a line. Choosing 3 and 10 we are forced to have {3,10,18,26} as a line. The first two points of the next three lines force the remaining points: {4,11,17,22}, {5,12,15,25}, {6,13,16,19}.

This leaves 7 and 8 with 20 and 21, while 20 and 21 are on a 7-point line. So S does not exist.

Suppose $n = 4$. In addition to the above possibilities for r_p, we have $r_p = n - 1$, all lines being $(n + 1)$-lines.

If L is a unique 5-line, then each point not on L is on a unique 3-line missing L, thus partitioning S into disjoint subsets of order 3 with one of order 5, while this is impossible since $v = 13$. Suppose S has at least two 5-lines. Since a 6-point is only on 3-lines, any two 5-lines meet each other.

Suppose finally that some point p has at least two, and therefore precisely two or three, 5-lines on it. If p has three 5-lines, then every other point is a 5-point on one 5-line and four 3-lines. Hence $S - p$ is the Nwankpa–Shrikhande plane or the affine plane of order 4 less one line and all its points. Suppose no point is a 3-point, and hence let p be a 4-point on the two 5-lines L and M. There can be no 5-lines not on p, and so points not on L or M are 6-points. Points of L and M distinct from $L \cap M$ are 5-points. Let 1 denote the point $L \cap M$ and 2, 3, 4, 5 the points not on L or M. We may assume that $\{1,2,3\}$ and $\{1,4,5\}$ are the 3-lines on 1. What other points then are on a line with 2 and 4? Certainly none of 1, 3, 5. Let 6 denote the third point on 24. So 6 is on L or M. But each line on 6 then meets both L and M and so 24 has four points, a contradiction.

Case 3. $v = n^2 - 2$. In this case, Proposition 3.3.2 yields $r_p = n + 1$ for all p, with $(n - 1)/2$ of these being $(n + 1)$-lines and $(n + 3)/2$ being $(n - 1)$-lines. Applying Proposition 3.1.1 implies

$$n + 1 | (n^2 - 2) \tfrac{1}{2} (n - 1) = \tfrac{1}{2} n^2 (n + 1) - (n + 1) n + 1,$$

and so $\frac{1}{2} (n + 1) | 1$, which is impossible.

Case 4. $v = n^2 - 1$ and $n \geq 23$. By Proposition 3.3.2, $r_p = n$, with $n - 1$ $(n + 1)$-lines and one $(n - 1)$-line on p, $r_p = n + 1$ with $n/2$ $(n + 1)$-lines and $n/2 + 1$ $(n - 1)$-lines on p, or $r_p = n + 2$ with one $(n + 1)$-line and $n + 1$ $(n - 1)$-lines on p.

If some point p is on *all* $(n + 1)$-lines, let L be an $(n - 1)$-line not on p. Each point of L is joined to p by an $(n + 1)$-line. Thus there is precisely one $(n - 1)$-line M on p, and each point of $M - p$ is on an $(n + 1)$-line not on p, which is a contradiction. So no n-points exist.

Suppose that the $(n + 1)$-line M misses the $(n - 1)$-line L. Then $r_p = n + 2$ for all $p \in L$, and so there are n $(n - 1)$-lines on each point of L meeting M. Let s be the number of points of M on precisely $n + 1$ lines. Then the number of $(n - 1)$-lines on points of M which meet L is both

$$\leq \left(\frac{n}{2} + 1\right) s + (n + 1 - s)(n - 1) \quad \text{and} \quad \geq \frac{ns}{2} + (n + 1 - s)(n - 1).$$

So $s + ns/2 + n^2 - 1 - sn + s \geq n^2 - n \geq ns/2 + n^2 - 1 - sn + s$, giving $s + 1 \geq (n/2 - 1)(s - 2) \geq 1$ and $n \leq 10$, a contradiction. Thus each $(n + 1)$-line meets each $(n - 1)$-line.

Assume now that all $(n + 1)$-lines meet each other. Then $r_p = n + 1$ for all p, and since all lines meet a fixed $(n + 1)$-line, $b = n^2 + n + 1$ where $(n + 2)(n + 1)/2$ of these are of degree $n - 1$ and $n(n - 1)/2$ are of degree $n + 1$. By Theorem 2.3.5, S is a projective plane of even order n less an $(n + 2)$-arc.

Now let L and M be disjoint $(n + 1)$-lines. So all points on L and M have $r_p = n + 2$. Let $p \notin L, M$. Then p is on at least $n + 1$ $(n - 1)$-lines, and so $r_p = n + 2$. Thus $r_p = n + 2$ for *all* p, and no two $(n + 1)$-lines meet. Thus there are $(n^2 - 1)/(n + 1) = n - 1$ $(n + 1)$-lines and $b = (n + 1)(n + 1) + n - 1 = n^2 + 3n$. By P. de Witte (1983), S is a projective plane of order $n + 1$ less three concurrent lines and all their points.

Case 5. $v = n^2$. By Proposition 3.3.2, $r_p = n + 1$ for all p, where $(n + 1)/2$ of these are $(n + 1)$-lines and $(n + 1)/2$ are $(n - 1)$-lines. Applying Proposition 3.1.1 we obtain $n - 1 | n^2 \frac{1}{2}(n + 1)$, while $(n, n - 1) = 1$ if $n \geq 2$ and $(\frac{1}{2}(n + 1), n - 1) = 1$ if $n \geq 4$. □

The remainder of this section finishes the proof of Theorem 3.3.1. We assume now that lines of each length $n - 1, n, n + 1$ exist. Note that Proposition 3.3.2 tells us $v \geq n^2 - n - 1$.

Proposition 3.3.4. If S is a linear space with $v \leq n^2$ points, $n \geq 23$, and line range $\{n - 1, n, n + 1\}$ with all three line sizes occurring, and if S has an $(n + 2)$-point, then $v = n^2 - 1$ and S is a projective plane of order $n + 1$ less three concurrent lines and all their points or $v = n^2$ and S is a projective plane of order $n + 1$ less three concurrent lines and all their points but one, not the point of intersection.

PROOF. Clearly, there can be no point of degree greater than $n + 2$. If there is an $(n + 2)$-point, then by Proposition 3.3.2, $v \geq n^2 - 3$.

If $v = n^2 - 3$, it follows that all lines through an $(n + 2)$-point p have degree $n - 1$.

For any fixed $(n + 1)$-line L, there is a line M on p parallel to L, which therefore contains only $(n + 2)$-points. Suppose there is an $(n + 2)$-point q not on M. Then q is also on a line M' parallel to L, whose points are all $(n + 2)$-points; furthermore, $M \cap M' = \phi$, as otherwise there is a point on at least $n + 3$ lines. Those lines meeting and distinct from M all meet L, and there are $(n - 1)(n + 1) = n^2 - 1$ of them. Only $(n - 1)(n - 1)$ of these meet M', leaving $2n - 2$ lines on L missing M'. But there are also $n^2 - 1$ lines meeting L and M' so that the number of lines meeting L is at least $n^2 - 1 + 2n - 2 + 1 = n^2 + 2n - 2$, while L is on at most $n(n + 1) + 1 = n^2 + n + 1$ lines. This implies $n^2 + 2n - 2 \leq n^2 + n + 1$ or $n \leq 3$, and a contradiction. Hence no such point q exists.

Let s be any point not on M. Then s is on at least three $(n - 1)$-lines, implying $r_s = n + 1$.

The linear space $S - M$ contains an n-line N. Since for all points s of $S - M$, $r_s = n + 1$, we get a partition of $S - M$ into $(n - 2)$-, $(n - 1)$- and $(n + 1)$-lines, noting that all $(n + 1)$-lines meet N. As $(n + 1)$-lines exist, the partition contains at least $n + 1$ lines, and so the number of points in $S - M$ is at least $n + n(n - 2) = n^2 - n$. Since the precise number is $n^2 - 3 - (n - 1) = n^2 - n - 2$, we have a contradiction.

Suppose now that $v = n^2 - 2$. By Proposition 3.3.2, $r_p = n$, and p is on $n - 2$ lines of degree $n + 1$, one line of degree n and one line of degree $n - 1$; $r_p = n$, and p is on $n - 3$ lines of degree $n + 1$ and 3 lines of degree n; $r_p = n + 1$, and p is on s_p lines of degree n, $(n - s_p - 1)/2$ lines of degree $n + 1$ and $(n - s_p + 3)/2$ lines of degree $n - 1$ where s_p is a non-negative integer; or $r_p = n + 2$, and p is on one line of degree n and $n + 1$ lines of degree $n - 1$.

Let L be a fixed $(n + 1)$-line, p an $(n + 2)$-point not on L, and M the line on p missing L. If M is an n-line, then there are $n(n + 1)$ $(n - 1)$-lines meeting M and L. So $n(n + 1) \leq (n + 1) \cdot \max\{1, \frac{1}{2}(n - \bar{s} + 3)\}$ where $\bar{s}$ is the minimum value of the numbers s_x of n-lines through an $(n + 1)$-point x, where x ranges over the $(n + 1)$-points of L. Clearly, therefore, $2n \leq n - \bar{s} + 3$ or $n - 3 \leq -\bar{s}$. This is impossible if $n \geq 4$. Hence M is in fact an $(n - 1)$-line.

Suppose there is an n-point s not on M. Then at most one line on s misses M and since no $(n + 1)$-lines meet M, we must have $n \leq 4$, a contradiction.

Letting L and M be as above, we now count the number of $(n - 1)$-lines meeting both L and M. There are precisely $n(n - 1)$ such lines on M meeting L. Therefore

$$n^2 - n \leq (n + 1)\tfrac{1}{2}(n - \bar{s} + 3),$$

where $\bar{s}$ is the minimum value of s_x as x ranges over the points of L. So

$$2n^2 - 2n \leq n^2 + 4n + 3 - \bar{s}(n + 1) \quad \text{or}$$

$$n^2 - 6n - 3 \leq -\bar{s}(n + 1).$$

For $n > 6$, this gives a contradiction.

Suppose then that $v = n^2 - 1$. Using Proposition 3.3.2, any point p is of one of the following types: (1) $r_p = n$, and p is on $n - 2$ lines of degree $n + 1$ and two lines of degree n; (2) $r_p = n$, and p is on $n - 1$ lines of degree $n + 1$ and one line of degree $n - 1$; (3) $r_p = n + 1$, and p is on s_p lines of degree n, $(n - s_p + 2)/2$ lines of degree $n + 1$, and $(n - s_p + 2)/2$ lines of degree $n - 1$; (4) $r_p = n + 2$, and p is on one line of degree $n + 1$ and $n + 1$ lines of degree $n - 1$; (5) $r_p = n + 2$, and p is on two lines of degree n and n lines of degree $n - 1$.

Suppose there are two disjoint $(n + 1)$-lines L and M. So all points of L and M have degree $n + 2$ and are of type 4. Let $p \notin L, M$. Then p is also of type 4, as it is on at least $n + 1$ lines of degree $n - 1$. We therefore have a partition of **S** into $n - 1$ $(n + 1)$-lines, and all other lines are $(n - 1)$-lines. By Proposition 3.3.3, S is a projective plane of order $n + 1$ less three concurrent lines and all their points. We therefore assume that all $(n + 1)$-lines meet each other.

We shall prove that all lines meet any $(n + 1)$-line.

Suppose first that there is an n-line N missing an $(n + 1)$-line L. Points of N must be of the fifth type, so that any point p of L is on at most two $(n + 1)$-lines. If p is an $(n + 1)$-point, p must be on at least $n - 4$ and at most $n - 2$ lines of degree n, all meeting N. There are precisely n lines of degree n meeting N and L, and no $(n + 2)$-point of L is on n-lines, so there is a second $(n + 1)$-point of L also on at least $n - 4$ n-lines meeting N. This gives $2(n - 4) \leq n$ or $n \leq 8$, a contradiction. So all points of L must be of the fourth type, but as mentioned, this means no n-lines meet L, a contradiction.

Now suppose that N is an $(n - 1)$-line missing the $(n + 1)$-line L. Let a be the number of points of N of type 4. Let c be the number of $(n + 1)$-points of L. Thus the number of $(n + 1)$-lines on the $(n + 1)$-points of L is at least $a + 1$ and at most $a + c + 1$. So the number of $(n - 1)$-lines on these $(n + 1)$-points is at least $a + 2c$ and at most $a + 3c$. Hence, the number of n-lines on these points is at most

$$cn + 1 - [a + 1 + (a + 2c)] = cn - 2a - 2c.$$

Now the number of n-lines meeting N (all of which meet L) is $2(n - 1 - a)$ and therefore

$$cn - 2a - 2c \geq 2n - 2 - 2a,$$

$$c(n - 2) \geq 2n - 2 = 2(n - 2) + 2,$$

$$c > 2.$$

Also, the number of $(n - 1)$-lines on L is at most $(n + 1 - c)(n + 1) + a + 3c$, while the number on N meeting L is $an + (n - 1 - a)(n - 1)$. Therefore,

$$n^2 + 2n + 1 - cn + a + 2c \geq n^2 - 2n + 1 + a, 4n \geq c(n - 2),$$

and so $c \leq 4$. Hence $c = 3$ or 4.

Now the number of n-lines on N meeting L is $2(n - 1 - a)$. Let d be the number of n-lines on $(n + 1)$-points of L which do not meet N. So $cn - 2(n - 1 - a) - d$ of the lines on the $(n + 1)$-points of L are either $(n + 1)$-lines distinct from L, or $(n - 1)$-lines. So $\frac{1}{2}(cn + 1 - 2(n - 1 - a) - d - 2c)$ of these are $(n + 1)$-lines, and at least $\frac{1}{2}(cn + 1 - 2(n - 1 - a) - d - 2c) - c$ of them meet N. So

$$cn + 3 - 2n + 2a - d - 4c \leq 2a,$$

$$n(c - 2) < 4c + d - 3,$$

$$n \leq 4 + (d + 5)/(c - 2) \leq 12,$$

a contradiction.

Thus *all* lines meet $(n + 1)$-lines. Suppose there is an n-point p; then it is unique. If p is of type 1, then setting $n = m + 1$, $S - p$ is a linear space with $m^2 + m - 1$ points and line sizes m and $m + 1$. By Proposition 3.2.7, $S - p$ does not exist. Assume p is of type 2. Let N be the $(n - 1)$-line on p, and $q \in N - p$. Any second line on q meets all $(n + 1)$-lines on p and hence can only be an n-line. Each point of $S - N$ is on at least $n - 2$ n-lines, and so is of type 2 with $s_p = n - 2$. By Theorem 3.2.3, $S - N$ is an affine plane of order n less a line and all its points but one. Therefore S is a projective plane of order n less a line and all its points but one, and less two additional points of a line on this remaining point; but such a space has no $(n + 2)$-point.

We assume now that there are no n-points. Clearly, any point not on a fixed $(n + 1)$-line has degree $n + 1$. Suppose some point q is on all $(n + 1)$-lines. If the number of $(n + 1)$-lines is more than one, then r_q is still $n + 1$. So suppose q is on a unique $(n + 1)$-line L. If $r_q = n + 2$, all other lines on it are $(n - 1)$-lines, and since each $p \notin L$ is on a unique $(n-1)$-line, this line is pq. So all other points of L are *only* on n-lines, which is a contradiction. Hence all points have degree $n+1$.

Finally, fix an n-line N. We get a partition into a n-lines and c $(n-1)$-lines, yielding $an + c(n-1) = n^2 - 1$. Thus $n|c - 1$ and $n - 1|a > 0$. So $a = n - 1$ and $c = 1$ and $a + c = n$, so that $(n+1)$-lines cannot exist, a contradiction.

Suppose lastly that $v = n^2$. By Proposition 3.3.2, any point is of one of the following types: (1) $r_p = n$, and p is on $n - 1$ lines of degree $n + 1$ and one line of degree n; (2) $r_p = n + 1$, and p is on s_p lines of degree n, $(n + 1 - s_p)/2$ lines of degree $n + 1$, and $(n + 1 - s_p)/2$ lines of degree $n - 1$, s_p a non-negative integer; (3) $r_p = n + 2$, and p is on 3 lines of degree n and $n - 1$ lines of degree $n - 1$; (4) $r_p = n + 2$, and p is on one line of degree $n + 1$, one line of degree n, and n lines of degree $n - 1$.

If an n-point p exists, then $S - p$ has only n- or $(n-1)$-lines. By Theorem 3.2.3, $S - p$ is a punctured affine plane of order n. Hence, S is a projective plane of order n less a line and all its points but one, and less an additional point. So we suppose that n-points do not exist.

Suppose there are two parallel $(n+1)$-lines L and M; so all points of L and M are of type 4 above. Let p be a point not on L or M. If it has degree $n + 1$, then it is only on n-lines. Since each point of L is on precisely one n-line, p is a unique $(n+1)$-point. Thus every other point is an $(n+2)$-point, and since it can be on at most two n-lines, it is of type 4 above. So p is on *all* n-lines, and $S - p$ has only $(n+1)$- and $(n-1)$-lines. By Proposition 3.3.3, S is a projective plane of order $n + 1$ less three concurrent lines and all their points but one, not the point of intersection.

Now suppose all points have degree $n + 2$. If some point is of type 3 above, we have an $(n-1)$- or an n-line N missing L, and either $2n$ or $3(n-1)$ lines of degree n meeting N and L, while L meets precisely $n + 1$ lines of degree n. This forces $n \leq 2$, and a contradiction. So all points are of type 4, and it follows that we have a partition of S into $(n+1)$-lines, implying $n + 1|n$, a contradiction.

Hence, we may assume that all $(n+1)$-lines meet each other.

Suppose there is an $(n+2)$-point not on the fixed $(n+1)$-line L. Hence it is on an n- or $(n-1)$-line N missing L. Let a be the number of points of type 4 on N, and c the number of $(n+1)$-points on L. There are precisely $a + 1$ lines of degree $n + 1$ on $(n+1)$-points of L. Since this includes L, there are precisely $a + c$ lines of degree $n - 1$ in all on these points. Therefore the number of n-lines on $(n+1)$-points of L is

$cn + 1 - (a + 1 + a + c) = cn - 2a - c$. The number of n-lines on N but not equal to N is either $2(n - a)$ or $a + 3(n - 1 - a)$, the smaller of which is $2(n - a)$. All of these lines meet L. Therefore,

$$2(n - a) \le cn - 2a - c + (n + 1 - c)$$
$$n - 1 \le c(n - 2),$$
$$1 < (n - 1)/(n - 2) \le c.$$

So $c \ge 2$.

Also, the number of $(n - 1)$-lines on N but distinct from N is either $na + (n - a)(n - 1)$ or $a(n - 1) + (n - 1 - a)(n - 2)$, the smaller of which is the latter. All of these lines meet L. The number of $(n - 1)$-lines on L is $a + c + (n + 1 - c)n$. So

$$n^2 - 3n + 2 + a \le n^2 + n + a + c - cn,$$
$$c(n - 1) \le 4n - 2 = 4(n - 1) + 8,$$

and so $c \le 4$.

Now suppose that there are (at least) two $(n + 1)$-points q, $q' \notin L$. There are (at least) $2(n + 1 - c) - 1$ distinct lines joining q and q' to the $(n + 2)$-points of L. At most $n + 1 - c$ of these can be n-lines and the rest, at least $2(n + 1 - c) - 1 - (n + 1 - c) = n - c$, are $(n - 1)$-lines. Since any $(n + 1)$-point has the same number of $(n + 1)$-lines as $(n - 1)$-lines, q and q' are on at least $n - c - 1$ $(n + 1)$-lines. So q and q' are on at least $2(n + 1 - c) - 1 + n - c - 1 = 3(n - c)$ lines, while they are on at most $(n + 1) + n = 2n + 1$. So $3(n - c) \le 2n + 1$ or $n \le 3c + 1 \le 13$, a contradiction. It follows that there is at most one $(n + 1)$-point not on L.

Suppose s is an $(n + 1)$-point not on L, and p and q are $(n + 1)$-points of L. If s is only on n-lines, then $S - s$ has $n^2 - 1$ points, and $(n - 1)$-, n- and $(n + 1)$-lines, and we may apply the last case. Since there are precisely c $(n + 1)$-points of L, s is on at most c $(n + 1)$-lines. Each point of $S - s$ is on a line parallel to L, giving a partition P of $(S - L) \cup \{s\}$ into n- and $(n - 1)$-lines. (Recall that all $(n + 1)$-lines meet each other.) Letting a be the number of n-lines and d the number of $(n - 1)$-lines in P, we obtain

$$n + 1 + an + d\,(n - 1) = n^2 - 1,\ an + d\,(n - 1) = n^2 - n - 2,$$

so $n | d - 2$, $n - 1 | a + 2 > 0$. Thus $a = n - 3$ and $d = 2$. So including L, there are n lines in P. This means that all $(n + 1)$-lines except L meet s.

The number of n-lines meeting L now depends on the number of $(n + 1)$-lines as follows:

one $(n + 1)$-line on s $\quad c(n - 1) - 2 + (n + 1 - c) = (c + 1)n - 2c - 1$

two $(n + 1)$-lines on s $\quad c(n - 1) - 2 \cdot 2 + (n + 1 - c) = (c + 1)n - 2c - 3$

three $(n + 1)$-lines on s $\quad c(n - 1) - 2 \cdot 3 + (n + 1 - c) = (c + 1)n - 2c - 5$

four $(n + 1)$-lines on s $\quad c(n - 1) - 2 \cdot 4 + (n + 1 - c) = (c + 1)n - 2c - 7$

All lines on s meet L. So L is on $c(n - 2)$ n-lines missing s and therefore meeting all lines of the partition P. Let N be an n-line of P, and c' the number of $(n + 1)$-lines on s. There are $2(n - c')$ n-lines meeting N, distinct from N, and at least $n + 1 - 2c' - 1 = n - 2c'$ of these pass through s. So at most n n-lines meet N, while $c(n - 2) = n + [(c - 1)n - 2] > n$, giving a contradiction. Thus, s does not exist.

Suppose then, that p and q on L are the only $(n + 1)$-points. Again we get a partition into a n-lines and d $(n - 1)$-lines where

$$n + 1 + an + d(n - 1) = n^2, \; an + d(n - 1) = n^2 - n - 1$$

so that $n | d - 1$, $n - 1 | a + 1 > 0$. Thus $a = n - 2$ and $d = 1$. So there are n lines again in this partition P. Thus L is a unique $(n + 1)$-line and any n-line not in P meets every line of P. The number of n-lines meeting L is $c(n - 1) + n + 1 - c = (c + 1)n - 2c + 1$. The number meeting an n-line of P, and not equal to L is $2n$, so we have the contradiction $n \leq 7$.

We may therefore suppose that *no* point not on L is an $(n + 2)$-point. Suppose some point p of L is of type 4. Let N be an $(n - 1)$-line on p, and q a point of $N - p$. So q is on an $(n + 1)$-line $M \neq L$, $p \notin M$. Since $r_p = n + 2$, there is a line N' on p missing M, and $N' \neq L$. But then each point of N' is on precisely $n + 2$ lines, a contradiction. □

Proposition 3.3.5. Let S be a linear space with $v \leq n^2$, $n \geq 23$, all points of degree $n + 1$, and line range $\{n - 1, n, n + 1\}$ with all three line sizes occurring. Then S is a projective plane of order n less an $(n + 1)$-arc.

PROOF. Clearly $b = n^2 + n + 1$.

Fix an n-line N. Any point not on N is on a unique line missing N. We thus obtain a parallel class of $n + 1$ lines, and counting v in two ways yields the information that the parallel class has one n-line and n $(n - 1)$-lines and that $v = n^2$. Adjoining a corresponding 'new point' p to S, we obtain a linear space with $n^2 + 1$ points $n^2 + n + 1$ lines, constant point size $n + 1$ and line range $\{n - 1, n, n + 1\}$. The point p is on n n-lines, each of which can be used to construct a different 'new point' as above. This produces a linear space with $n^2 + n + 1$ points and $n^2 + n + 1$ lines, which must therefore be a projective plane of order n. It follows that S must be as in the proposition. □

Proposition 3.3.6. Let S be a linear space with $v \leq n^2$, $n \geq 4$, no $(n + 2)$-points, and line range $\{n - 1, n, n + 1\}$ with all three line sizes occurring. Suppose further that S has n-points. Then S is a projective plane of order n less a line and all its points but one and perhaps also less two points of a second line on this point while retaining the point, or less two lines and all their points except one point on each line, not the point of intersection.

PROOF. Any n-point is on all $(n + 1)$-lines. Thus if there is more than one n-point, there is a unique $(n + 1)$-line, and it is on all n-points. Since all lines meet an $(n + 1)$-line we obtain $b \leq n^2 + n + 1$. Any n-line M determines a partial parallel class. Letting s be the number of n-points, we obtain $b = n^2 + n + 2 - s$, $1 \leq s \leq n + 1$, with respect to this class. On the other hand, $v \geq n^2 - n + 1$, and each point not on $L \cup M$, L a fixed $(n + 1)$-line, is on a unique line missing M but meeting L. This implies $(n + 1 - s)\, n + s \geq v \geq n^2 - n + 1$ and so $s \leq 2$. Therefore there are either one or two n-points. Let p be an n-point, and q the possible second n-point. It follows that

$$n^2 + n \leq b \leq n^2 + n + 1.$$

Clearly, any n-line determines a partial parallel class on $n + 1$, n or $n - 1$ lines, with p and q as the only points possibly not covered.

Fix an $(n - 1)$-line H. If H misses some n-line M, then H is in the above partial spread determined by M. All points of M, except perhaps one, are on a second line missing H. Suppose two of this set of lines meet in some point. Then this point is not an n-point and so it is on an element of the partial parallel class determined by M, and this has at least $n + 2$ lines through it, a contradiction. So the set of lines meeting M and missing H, along with H itself, forms a second partial parallel class on H with either $n + 1$ or n lines. Clearly H is on at most two such maximal partial parallel classes.

Suppose then, that the $(n - 1)$-line H meets all n-lines.

If p is the unique n-point, then $b = n^2 + n$. Let M be an n-line meeting

H and count the number of lines missing both M and H. This number is either 1 or 2 depending on where p lies with respect to M and H. Thus at least one line, say H', necessarily an $(n-1)$-line, misses both M and H in this case. As above, H' is in the unique maximal partial parallel class generated by M, and H' is in the second maximal partial parallel class on H. This maximal partial parallel class determined by H and H' has either n or $n+1$ lines depending on the position of p relative to M and H. We consider three cases.

If $p \in M \cap H$, then $v = n^2 - 1$. On the other hand, M determines a parallel class covering at least n^2 points, a contradiction. If $p \in M$, $P \notin H$, then $v = n^2 - 1$ or $n^2 - n + 1$. However, once again M determines a parallel class covering at least n^2 points. Finally, if $p \notin M$, $p \in H$, we have $v = n^2 - 1$. We may in this case assume, without loss of generality, that no n-line is on p. Hence p is on a unique $(n-1)$-line, and on $n-1$ lines of degree $n+1$. Then all n-lines meet the $(n-1)$-line on p, which we shall now call H, and any line meeting H not in p meets all lines on p and is therefore an n-line. Thus there are $n(n-2)$ n-lines; moreover, each n-line determines a parallel class of $S - p$ with $n-2$ n-lines and $2\ n -$ 1-lines. There are therefore n such parallel classes. Introducing a new system with point set $p^* = p \cup \{n$ parallel classes above$\}$ and line set $\mathcal{L}^*$ comprising lines of S, with each line not on p getting one new point corresponding to a parallel class, along with one additional line consisting of p along with all n parallel classes, $S^* = (p^*, \mathcal{L}^*)$ is a linear space with $n^2 + n - 1$ points in which each point is on $n + 1$ lines. S^* is a projective plane of order n less two points. It follows that S is a projective plane less one line and all its points but one, and less two additional points on a second line on this point.

Suppose then that H meets all n-lines, and p and q are both n-points. Then $L = pq$ is the unique $(n+1)$-line and $b = n^2 + n - 1$. If p is on no n-line, we obtain the contradiction $v = n^2 - 2n + 3$; so let M be a fixed n-line on p. Counting the number of lines missing both M and H we obtain the numbers 1 or 2 depending on the position on p and q relative to H. So in any case, we may assume that there is an $(n-1)$-line H' missing both M and H. As above, H' is in the unique maximal partial parallel class determined by M, and H' is in the second partial parallel class on H. As before, we consider three cases: $p \in M \cap H$; $p \in M$, $p \notin H$; $p, q \notin M, H$. Counting points on lines of the maximal partial parallel class on H and H' we see in the first and third cases $v = n^2 - 1$ immediately. The second case gives $v = n^2 - n$, $n^2 - n + 1$ or $n^2 - 1$. However, M determines a maximal partial parallel class covering at least $n + 1 + (n - 1)^2$ points, so that $v = n^2 - 1$ is the only possibility. We eliminate the first two cases now as follows. Since $n^2 - n + 2 < n^2 - 1$, there is an n-line $\bar{M}$ missing M and not on p. But this is impossible, since p is an n-point. We may also conclude that neither p nor q is on an n-line. Then counting points on lines through p, we obtain $v = n + 1 + (n - 1)(n - 2) = n^2 - 2n + 3$, a contradiction.

We are able to assume then, each $(n - 1)$-line is in precisely two maximal partial parallel classes of S. Since corresponding to each n-line there is a unique maximal partial parallel class we are able to introduce a new structure with point set $p^* = p \cup \{$the maximal partial parallel classes determined by n-lines$\}$ and line set $\mathcal{L}^*$ comprising the lines of $\mathcal{L}$ along with any points of p^* to which they belong, together with two new lines defined as described below. Clearly $v \geq (n + 1) + (n - 1)^2 = n^2 - n + 2$ if there is an n-line not on p or q, and $v \geq n + 1 + n(n - 2) = n^2 - n + 1$ if all n-lines are on p or q. But counting the number a of n-lines on p (respectively q), we obtain $v = n + 1 + a(n - 1) + (n - 1 - a)(n - 2) \geq n^2 - n + 1$, or $n - 1 \geq a \geq n - 2$. Let p along with all maximal partial parallel classes corresponding to n-lines on q be a new line. Similarly, let q along with all maximal partial parallel classes corresponding to n-lines on p be a second new line. It is not difficult to check that the new structure $S^* = (p^*, \mathcal{L}^*)$ is a linear space.

If all n-lines of S are on p or q, then each is on precisely one $(n - 1)$-line. But an $(n - 1)$-line misses some n-line, which is impossible here. If there is an n-line of S not on p or q, then since each $(n - 1)$-line misses some n-line, every line of S^* has either n or $n + 1$ points. Moreover, $v^* = n^2 + n$ and $b^* = n^2 + n + 1$, so that S^* is a partial projective plane of order n. It follows that S is a projective plane of order n less two lines and all their points except one point on each line, not the point of intersection. □

3.4 Two non-consecutive line sizes

In Theorem 2.3.7, we saw that a linear space with line degrees $m^2 - 1$ and $m^2 - m - 1$ was, under certain other conditions, embeddable in a projective plane P of order m^2. In fact, it is the complement of the union of two disjoint Baer subplanes of P.

A. Beutelspacher (1986) considers the case of two line sizes n and $n - k$ where $n \geq 2k + 3$, $(k, n - 1) = (k + 1, n) = 1$, $b \leq n^2 + n + 1$, and where there exists a line of degree n. (In particular if $n = ak$ where $(k + 1, a) = 1$, and even more particularly, if n is a power of k, then

$(k, n - 1) = (k + 1, n) = 1$ is satisfied.) His classification of such linear spaces gives a 'near-embeddability' result:

Theorem 3.4.1. Let S be a non-trivial finite linear space in which any line has n or n − k points, $k \geq 1$ where $n \geq 2k + 3$ and $(k, n - 1) = (k + 1, n) = 1$. Furthermore, suppose that $b \leq n^2 + n + 1$ and that there is at least one n-line. Then S is one of:

(i) a (punctured) projective plane of order $n - 1$;
(ii) an affine plane of order n;
(iii) an affine plane of order n less all points but one of a suitable line, along with the line itself;
(iv) the complement of a set of points in a projective plane of order n, such that each line of s meets the set in 1 or $k + 1$ points;
(v) the pseudo-complement of a k-pencil in a projective plane of order $n - 1$;
(vi) the pseudo-complement of a $(k + 1)$-pencil in a projective plane of order n;
(vii) the pseudo-complement of a maximal k-arc in a projective plane of order n which has an $(n + 1)$-line deleted.

Unless otherwise stated, the results of this section pertain to linear spaces for which the assumptions of Theorem 3.4.1 hold.

Proposition 3.4.2. If all lines of S are n-lines, then S is a projective plane of order $n - 1$ or an affine plane of order n.

PROOF. This follows immediately from Theorem 3.2.2. □

Proposition 3.4.3. If there exists a point p with $r_p < n$, then $S - p$ is the pseudo-complement of a k-pencil in a projective plane of order $n - 1$.

PROOF. Clearly, all n-lines are on p. Suppose there is an $(n - k)$-line on p. Let q be a point of an n-line and s be a point on no n-line. Then

$$v - 1 = n - 1 + (r_q - 1)(n - k - 1),$$

while $v - 1 = r_s(n - k - 1)$.

Hence $n - k - 1 | n - 1$, contradicting $(k, n - 1) = 1$.

Therefore any line on p is an n-line. So $v - 1 = r_p (n - 1)$ together with the first equation above in $v - 1$ implies $n - k - 1 | r_p - 1$. Write $r_p - 1 = a(n - k - 1)$, which is $< n - 1$. Thus $(a - 1)n < a(k + 1) - 1$, and since $n \geq 2k + 1$ this forces $a = 1$ and $r_p - 1 = n - k - 1$. Consequently $v = (n - k)(n - 1) + 1$.

Setting $v - 1 = n - 1 + (r_q - 1)(n - k - 1) = (n - k)(n - 1)$ we obtain $r_q = n$ for all $q \neq p$. It follows that $b = r_p + (n - 1)^2 = (n - 1)^2 + (n - 1) + 1 - k$. Thus $S - p$ is the pseudo-complement of a *k-pencil* in a projective plane of order $n - 1$. □

Proposition 3.4.4. If not all points of S have the same degree, then either (iii) or (v) of Theorem 3.4.1 holds.

PROOF. In view of Proposition 3.4.3, we may assume that $r_p \geq n$ for all points p.

Step 1. We prove first that any n-line has at least one point on at most $n + 1$ lines. Suppose L is an n-line for which this is not true. That is, $r_p \geq n + 2$ for all $p \in L$. Then $n^2 + n + 1 \geq b \geq (n + 1)n + 1$ implies that all lines meet L, $r_p = n + 2$ for all points of L, $r_p = n$ for all points not in L. Suppose L' is a second n-line. Then any point of $L - L'$ is on two lines missing L', while no point of $S - (L \cup L')$ is on a line missing L'. Thus we have a contradiction, and conclude that L is unique. Finally, counting $v - 1$ using a point p of L and a point q outside L we obtain

$$\begin{aligned} n - 1 + (n + 1)(n - k - 1) &= n - 1 + (r_p - 1)(n - k - 1) \\ &= v - 1 = r_q(n - k - 1) = n(n - k - 1), \end{aligned}$$

and so $2n = k + 2$, a contradiction.

Step 2. We now show that $r_p \in \{n, n + k\}$ for all p.

Choose a point p such that r_p is minimal. By step 1, $r_p = n$ or $n + 1$. Without loss of generality, we may now assume that there is a point q with $r_q > r_p$. Letting a, respectively c, be the number of n-lines on p, respectively q, we get

$$ak + r_p(n - k - 1) = v - 1 = ck + r_q(n - k - 1)$$

whence

$$(r_p - r_q)(n - k - 1) = (c - a)k, \tag{1}$$

and so $n - k - 1 | a - c > 0$.

Then either $a - c = n - k - 1$ or $2(n - k - 1) \leq a - c$. If the inequality holds, then $2(n-k-1) \leq a - c \leq a \leq r_p \leq n+1$ gives $n \leq 2k+3$, and therefore $n = 2k+3$. It follows that $c = 0$ and $a = n+1 = r_p$. However, it now follows that the line pq is both an n-line and not an n-line. Thus we conclude that $a - c = n - k - 1$. Substituting in equation (1) gives $r_q = r_p + k$. Since this is true for any q with $r_q > r_p$, it follows that there are only two point degrees possible and $r_p \in \{n, n+k\}$ for all p, or $r_p \in \{n+1, n+k+1\}$ for all p.

Assume that $r_p \in \{n+1, n+k+1\}$ for all p.

Fix an n-line L and let a be the number of $(n+k+1)$-points on it. Then $a(n+k) + (n-a)n = n^2 + ak$ lines meet L, excluding L itself. By step 1, $a < n$. Let p be an $(n+1)$-point on L, and let c be the number of n-lines on it. Choosing a point q of degree $n+k+1$ and letting d be the number of n-lines on it, we get, using the result $c - d = n - k - 1$ of step 2,

$$c = n - k - 1 + d \geq n - k - 1 \geq k + 2 \geq 2.$$

So there is an n-line $L' \neq L$ on p. On any point of L' there is at least one line missing L. Hence

$$b \geq n^2 + ak + n.$$

It follows that $ak \leq 1$ and so $a = 0$ or 1. If $a = 1$, then $k = 1$ and this leads to a contradiction by Exercise 3.6.8. So $a = 0$. Thus all points of an n-line are $(n+1)$-points. It follows that $d = 0$ and

$$v = (n+k+1)(n-k-1) + 1 = n^2 - k^2 - 2k.$$

Now using Proposition 3.1.1, we obtain

$$nb_n = cv_{n+1} + dv_{n+1+k} = cv_{n+1} = (n-k-1)v_{n+1}.$$

Thus $n - k - 1 | b_n$.

The number of n-lines meeting L in precisely one point is $n(n-k-2)$. On the other hand, any n-line missing L meets L', so there are at most n n-lines missing L. Therefore

$$(n-2)(n-k-1) < n(n-k-2) + 1 \leq b_n$$
$$\leq n(n-k-2) + n = n(n-k-1).$$

It follows that $b_n = (n-1)(n-k-1)$ or $n(n-k-1)$.

If $b_n = n(n-k-1)$, substituting above yields $v_{n+l} = n^2 > v$, a contradiction. Thus $b_n = (n-1)(n-k-1)$, from which it follows that $v_{n+l} = n(n-1)$ and $v_{n+k+l} = n - k^2 - 2k$.

Now the number of $(n - k)$-lines intersecting L is given by $n(n + 1 - (n - k - 1)) = n(k + 2)$. So the number s of $(n - k)$-lines missing L satisfies $s \leq n^2 + n + 1 - (n - 1)(n - k - 1) - n(k + 2) = n - k$. On the other hand, there are exactly $b_n - n(n - k - 2) - 1 = k$ n-lines missing L. Since these latter lines are disjoint, they cover precisely k points of $L' - L$. Therefore any line missing L through one of the remaining $n - (k + 1)$ points is an $(n - k)$-line. In particular, there are at least $n - k - 1$ $(n - k)$-lines missing L.

The partial parallel class of S formed by the $k + 1$ lines of degree n above, including L itself, covers $(k + 1)n$ points, all of which are $(n + 1)$-points. Consequently, the number v' of $(n + 1)$-points through which there is no n-line missing L is given by

$$v' = v_{n+1} - (k + 1)n = n(n - 1) - n(k + 1) = n(n - k - 2).$$

Through any one of these v' points there is precisely one line of degree $n - k$ missing L; while through any $(n + k + 1)$-point, there are $k + 1$ lines of degree $n - k$ missing L. Hence

$$s(n - k) = n(n - k - 2) + (n - k^2 - 2k)(k + 1).$$

This implies

$$2n = k^2 + 4k + 3 \quad \text{if } s = n - k - 1,$$

and

$$n(2k - 1) = k^3 + 4k^2 + 2k \quad \text{if } s = n - k.$$

From above, $v_{n+k+1} = n - k^2 - 2k \geq 0$ implies $n \geq k^2 + 2k$. Thus $s = n - k - 1$ implies $k = 1$ and $n = 4$ contradicting $n \geq 2k + 3$. On the other hand, if $s = n - k$, we obtain from $n \geq k^2 + 2k$ that $k \leq 2$. If $k = 2$ we would have the contradiction $3n = 8 + 16 + 4 = 28$. Thus $k = 1$ and $n = 7$, $v = 46$. By Theorem 3.2.1, this is impossible.

Step 3. We proceed to show that k must be 1.

Assume to the contrary that $k \geq 2$. By step 2, any point has degree n or $n + k$. Any line through an n-point intersects any n-line. By step 1, any n-line contains at least one n-point. Therefore any two n-lines meet each other.

Let the number of $(n + k)$-points on a fixed n-line L be a. Let p be an n-point and q an $(n + k)$-point each on c and d n-lines respectively. Then $b_n = 1 + a(d - 1) + (n - a)(c - 1) = 1 + n(c - 1) - a(n - k - 1)$. It follows that a is independent of the choice of L. Note also that any line disjoint from L necessarily is an $(n - k)$-line and contains only $(n + k)$-points.

Consider the case of an $(n - k)$-line M meeting any n-line. Let e be the number of $(n + k)$-points on M. Then

$$b_n = ed + (n - k - e)c = (n - k)c - e(n - k - 1).$$

Substituting in the above equation for b_n, we get

$$k(c - 1) = (a - e + 1)(n - k - 1).$$

Since $(k, n - 1) = 1$, it follows that $n - k - 1 | c - 1$. By the arguments of step 2, we also know that $c - d = n - k - 1$, so that $n - k - 1 | d - 1$. Assuming $d - 1 > 0$ we have

$$n - k - 1 \leq d - 1 = c - 1 - (n - k - 1) \leq r_p - 1 - (n - k - 1) = k,$$

contradicting $n \geq 2k + 3$. Hence $d = 1$, and so $c = n - k$ and

$$\begin{aligned} v &= 1 + r_q (n - k - 1) + dk \\ &= 1 + (n + k)(n - k - 1) + k = n^2 - k^2 - n + 1. \end{aligned}$$

Also, the first equation above in b_n yields

$$b_n = 1 + (n - a)(n - k - 1).$$

Since any n-line has precisely a points of degree $n + k$, and through any such point there is precisely one n-line, we have

$$v_{n+k} = b_n a = [1 + (n - a)(n - k - 1)]a.$$

Hence, $n^2 - k^2 - n + 1 = v \geq v_{n+k} = [1 + (n - a)(n - k - 1)]a \geq 1 + a(n - a)(n - k - 1)$, since $a \geq 1$.

By step 1, $a \leq n - 1$. If $a = n - 1$ then any n-line L has precisely one n-point and any n-line meeting L meets it in this point. Let s be a point on an $(n - k)$-line on this n-point.

Then s is on an n-line (using Proposition 3.2.1) which cannot meet L, a contradiction. So $a \leq n - 2$.

If $a \geq 2$, then $2 \leq a \leq n - 2$ implies $[a - (n - 2)](a - 2) \leq 0$, from which we obtain $a(n - a) \geq 2(n - 2)$. We also obtain as above,

$$\begin{aligned} n^2 - k^2 - n + 1 &\geq a + a(n - a)(n - k - 1) \\ &> 1 + 2(n - 2)(n - k - 1), \end{aligned}$$

which reduces to

$$k^2 + 4k + 4 < n(2k + 5 - n).$$

It follows, by our hypothesis, that $n = 2k + 3$ or $n = 2k + 4$. The latter conclusion yields $k(k + 2) < 0$, while the former yields $k = 1$. Each is a contradiction. Thus $a = 1$.

Now $2 \leq k + \dfrac{(a - e + 1)(n - k - 1)}{e - 1} = a - e + 1 = 2 - e$ implies $e = 0$. But by Proposition 3.2.1, $e = n - k$ and so $n = k$, a contradiction.

We conclude that an $(n - k)$-line always misses some n-line. This implies that any point of an $(n - k)$-line is an $(n + k)$-point, and that any line on an n-point is an n-line. Thus $c = n$, $d = k + 1$ and

$$b_n = 1 + n(n - 1) - a(n - k - 1)$$

from our initial equation in b_n. Any n-point lies on one of the $k + 1$ n-lines on the $(n + k)$-point q. Since any one of these lines is incident with exactly $n - a$ points of degree n, we have $v_n = (k + 1)(n - a)$. Thus $v_n n = v_n c = b_n(n - a)$ implies $b_n = n(k + 1)$, and so,

$$a(n - k - 1) = n(n - k - 1) - (n - 1),$$

implying $n - k - 1 | n - 1$, a final contradiction.

Step 4. Here, $k = 1$, and so the line degrees are n and $n - 1$, and the point degrees are n and $n + 1$. By considering an n-point q and an $(n + 1)$-point p, we see that

$$(n + 1)(n - 2) = r_p(n - 2) \leq v - 1 \leq r_q (n - 1) = n(n - 1).$$

Therefore

$$(n - 1)^2 + (n - 1) - 1 \leq v \leq (n - 1)^2 + (n - 1) + 1.$$

Since $n \geq 2k + 3$, we have $n \geq 5$. By Theorem 3.2.3, S is an affine plane of order n less a line and all its points but one. □

Proposition 3.4.5. Suppose all points of S have the same degree r. Then $r = n$ and S is as in (i), or $r = n + 1$ and S is one of (ii), (iv), (vi) or (vii) of Theorem 3.4.1.

PROOF. By Proposition 3.4.3, $r \geq n$. By step 1 of Proposition 3.4.4., $r \leq n + 1$.

Step 1. Suppose any point is incident with precisely n lines. Thus any line meets an n-line. So $b = 1 + n(n - 1) = n^2 - n - 1$. Let a be the (constant) number of n-lines through a point. Then

$$v = 1 + ak + n(n - k - 1) \text{ and } b_n = 1 + n(a - 1).$$

So

$$[1 + ak + n(n - k - 1)]a = va = nb_n = [1 + n(a - 1)]n.$$

The solutions to this quadratic in a are $a = n$ and $a = \frac{n-1}{k}$. If $a = n$, then all lines are n-lines, a contradiction. Since $(n - 1, k) = 1$, we conclude that $k = 1$ and $v = n(n - 1)$. Hence S is a punctured projective plane of order $n - 1$.

Step 2. Suppose any point is incident with precisely $n + 1$ lines. Clearly, for any fixed n-line L, the set of lines missing L, along with L itself, forms a parallel class.

Let a be the (constant) number of n-lines through any point. Then $v = 1 + ak + (n + 1)(n - k - 1)$.

Consider first of all the case $a = 1$. Thus $v = n(n - k)$. Moreover, $nb_n = an(n - k)$ implies $b_n = n - k$. It follows that any $(n - k)$-line meets every n-line and counting the $(n - k)$-lines which meet a fixed n-line yields $b_{n-k} = n^2$. So $b = n^2 + n + 1 - (k + 1)$ and S is the pseudo-complement of a $(k + 1)$-pencil in a projective plane of order n.

Assume now that $a > 1$. Thus there exists a pair of intersecting n-lines L and L'. Since the number of lines meeting one or both of these is at least $n^2 + n$, we conclude that $b \geq n^2 + n$.

Step 3. Suppose here that $b = n^2 + n$ and $a > 1$. From the argument in step 2, we see that there are precisely n lines in the parallel class of L. Moreover, no two distinct parallel classes have a common line.

Let the number of n-lines in a parallel class be c. Then $1 + ak + (n + 1)(n - k - 1) = v = cn + (n - c)(n - k) = ck + n(n - k)$, and so $c = a - 1$. Consequently

$$b_n = n(a - 1) + c = (n + 1)(a - 1).$$

The number of parallel classes is now $b_n/c = n + 1$. As no two parallel classes intersect, it follows from $b = (n + 1)n$ that S admits a parallelism. Adjoining $n + 1$ points at infinity corresponding to the $n + 1$ classes, and consequently a line at infinity on all $n + 1$ points, we obtain a linear space S^* with $n^2 + n + 1$ lines in which any point has degree $n + 1$, and each line has degree $n - k + 1$ or $n + 1$.

Finally, $[1 + ak + (n + 1)(n - k - 1)]\, a = va = nb_n = (n + 1)(a - 1)n$ reduces to $a^2k - a(kn + n + k) + n(n + 1) = 0$, implying $a = n + 1$ or $a = n/k$. The first is impossible. Hence $v = (n + 1)(n - k)$ and S^* has $(n + 1)(n + 1 - k)$ points. S^* is therefore the pseudo-complement of a maximal k-arc in a projective plane of order n.

Step 4. Suppose, lastly, that $b = n^2 + n + 1$ and $a > 1$. It follows from step 2 that each parallel class has precisely $n + 1$ elements, and any two distinct parallel classes intersect in exactly one line.

Let M be an $(n - k)$-line. There are $b_n - (n - k)a$ n-lines disjoint from M. Any such n-line determines a parallel class on M and so no n-line of this parallel class meets M. The number of parallel classes on M is therefore

$$\frac{b_n - (n - k)a}{c} = \frac{n(a - 1) + c - (n - k)a}{c} = 1 + k.$$

Define now the incidence structure $S^* = (p^*, \mathcal{L}^*)$ where p^* is p along with all parallel classes of S, and each line of $\mathcal{L}^*$ is the corresponding line of $\mathcal{L}$ with its points at infinity. Because any two parallel classes have precisely one common line, S* is a linear space. Also, each point and line has degree $n + 1$ in S^*. That is, S^* is a projective plane of order n by Proposition 1.2.8. It follows that S is as in (iv) of Theorem 3.4.1.□

Proposition 3.4.5 and so Theorem 3.4.1 are now completed.

3.5 On the existence of linear spaces with certain line sizes

In this last section of Chapter 3 we mention existence problems in connection with line sizes. This topic does not fit very comfortably with the first four sections of Chapter 3, but the results are important and should appear somewhere in this book.

R. M. Wilson developed a powerful asymptotic theory in the early 1970s which showed that once v is large enough, linear spaces with any combination of line sizes exist. We state his main theorem for linear spaces below without proof.

Let K be a set (finite or infinite) of positive integers. Denote by $B(K)$ the set of positive integers v for which there exists a linear space on v points with line sizes from K.

Define

$$\alpha(K) = \text{g.c.d.}\ (\{k - 1 | k \in K\})$$
$$\beta(K) = \text{g.c.d.}\ (\{k\,(k - 1) | k \in K\}).$$

Clearly, if $v \in B(K)$, then

$$v - 1 \equiv 0\ (\text{mod}\ \alpha(K)) \qquad \text{and} \qquad v\,(v - 1) \equiv 0\ (\text{mod}\ \beta(K)).$$

Theorem 3.5.1. (Wilson 1972, 1975.) For any set K of positive integers, B(K) contains all sufficiently large integers v satisfying the congruences $v - 1 \equiv 0$ (mod $\alpha(K)$) and $v(v - 1) \equiv 0$ (mod $\beta(K)$).

In 1983, D. A. Drake and J. A. Larson decided the existence problem of a linear space on v points with no lines of sizes 2, 3 or 6, for all v except $v = 30$. The case $v = 30$ is still open, but quite a lot is known about this situation.

The strange condition of line sizes not 2, 3 or 6 is explained by the following theorems.

Theorem 3.5.2. (Drake and Lenz 1975.) The existence of a non-trivial linear space on v points containing a k-line implies the existence of a self-orthogonal Latin square of order v with a subsquare of order k.

Theorem 3.5.3. (Brayton, Coppersmith and Hoffman 1976.) There is a self-orthogonal Latin square of order n if and only if n is a positive integer which is not a divisor of 6.

For the case $v = 30$, Drake and Larson have proved that only four line sizes are possible – 4, 5, 7 and 8 – and that each of these occurs respectively 41, 14, 1, 1; 22, 24, 3, 0; 37, 15, 3, 0; 24, 27, 1, 0; 29, 24, 1, 0; or 44, 15, 1, 0 times.

3.6 Exercises

1. Suppose S is a linear space with $v = n^2 + 2n + 4$ points, $n \geq 3$, in which each line has n points. Show that S is a (non-existent) 2-(19,3,1) design or a 2-(28,4,1) design.
2. If $v = n^2 + 2n + 2$, $n \geq 3$, and each line of S has either n or $n + 1$ points, show that S cannot exist.

3. If $v = n^2 + 2n + 3$, $n \geq 4$ and each line of S has either n or $n + 1$ points, show that S is one of the problematic 2-(51,6,1)-designs.
4. In Proposition 3.2.9, what happens if $n \leq 6$?
5. Suppose S has line degrees $n - 1$, n or $n + 1$, $4 \leq n \leq 6$ and $v = n^2 - 3$. If S is not one of the first eight types of Theorem 3.3.1, show that it must be of type (xi).
6. Show that there is no linear space with line degrees $n - 1$, n or $n + 1$, $n = 4$ and $v = n^2 - 2$.
7. Show that if $v = n^2$, $n \geq 8$ and S has line degrees $n - 1$, n or $n + 1$, then S is one of types (i) through (viii) of Theorem 3.3.1.
8. Prove the claim of step 2 of Proposition 3.4.4 that the case $a = 1$, $k = 1$ leads to a contradiction. (Hint: count v and apply Theorem 3.2.3.)

3.7 Research problems

1. Find a linear space with $v = 30$ and no 2-, 3- or 6-lines, or prove that such a space does not exist.

3.8 References

Batten, L. M. (1980), Linear spaces with line range $\{n - 1, n, n + 1\}$ and at most n^2 points. *J. Austral. Math. Soc.* (A) *30*, 215–228.

Batten, L. M. and Totten, J. (1980), On a class of linear spaces with two consecutive line degrees. *Ars Comb. 10*, 107–114.

Beutelspacher, A. (1986), Embedding linear spaces with two line degrees in finite projective planes. *J. Geometry 26*, 43–61.

Brayton, R. K., Coppersmith, D. and Hoffman, A. J. (1976), Self-orthogonal Latin squares. *Atti del Convegni Lincei 17*, 509–517.

Drake, D. A. and Larson, J. A. (1983), Pairwise balanced designs whose line sizes do not divide six. *J. Comb. Theory* (A) *34*, 266–300.

Drake, D. A. and Lenz, H. (1975), Finite Klingenberg planes. *Abh. Math. Sem.* Univ. Hamburg *44*, 70–83.

Hall, M. Jr. (1967, revised 1986), *Combinatorial Theory*, Blaisdell, Waltham, Mass.

Wilson, R. M. (1972a), An existence theory for pairwise balanced designs, I. Composition theorems and morphisms. *J. Comb. Theory* (A) *13*, 220–245.

Wilson, R. M. (1972b), An existence theory for pairwise balanced designs, II. The structure of PBD-closed sets and the existence conjectures. *J. Comb. Theory* (A) *13*, 246–273.

Wilson, R. M. (1975), An existence theory for pairwise balanced designs, III. Proof of the existence conjectures. *J. Comb. Theory* (A) *18*, 71–79.

de Witte, P. (1983). Variations on a theorem of Kuiper and Dembowski. *Simon Stevin 57*, 47–59.

de Witte, P. and Batten, L. M. (1983), Finite linear spaces with two consecutive line degrees. *Geom. Ded. 14*, 225–235.

4
Semiaffine linear spaces

4.1 *I*-Semiaffine linear spaces

If (p,L) is a non-incident point-line pair, we denote by $\pi(p,L)$ the number of lines through p which miss L. Using this notation, a linear space containing a quadrangle is a projective plane if and only if $\pi(p,L) = 0$ for every non-incident point-line pair. Likewise, a linear space containing a triangle is an affine plane if and only if $\pi(p,L) = 1$ for every non-incident point-line pair.

G. Pickert (1955) was the first to ask a natural generalization of these facts: What non-trivial linear spaces are characterized by

$$\pi(p,L) \leq 1$$

for every non-incident point-line pair?

In 1962, P. Dembowski gave a complete description of all such finite linear spaces, which he proceeded to call *semiaffine planes*. (Hence, the term 'semiaffinity condition' used in Section 1.4.) As P. Dembowski mentions in his 1962 paper, N. Kuiper had at about the same time, and independently, proved the same result. It is therefore now usually referred to as the Kuiper–Dembowski theorem. The aim of the first section of this chapter is to prove this result.

In subsequent sections we shall make yet a further generalization: Let I be a set of non-negative integers. S is said to be *I-semiaffine* if $\pi(p,L) \in I$ for every non-incident point-line pair (p,L) of S. S is said to be *I-affine* if S is I-semiaffine but not J-semiaffine for any proper subset J of I.

Clearly, any finite linear space is I-affine for a suitable set I. The semiaffine planes are of course the $\{0,1\}$-semiaffine linear spaces.

A number of people have already considered I-semiaffine sets for particular choices of I. M. Oehler (1975) characterized what he called *biaf-*

fine planes, which are simply our {1,2}-semiaffine linear spaces. A. Beutelspacher and N. Meinhardt (1984) introduced and characterized s-semiaffine linear spaces – our $\{s - 1,s\}$-semiaffine linear spaces. W. Hauptmann (1980) considered [0,2]-planes, which are our {0,1,2}-semiaffine linear spaces.

In Sections 4.3, 4.4 and 4.5, we shall study $\{0,s,t\}$-semiaffine linear spaces for s and t arbitrary positive integers with $s < t$.

4.2 {0,1}-Semiaffine linear spaces

In this section we prove the theorem due to N. Kuiper and to P. Dembowski.

Kuiper–Dembowski theorem. If S is a finite {0,1}-semiaffine linear space, then it is one of the following:

(a) a near-pencil,
(b) a projective or affine plane,
(c) a punctured projective plane,
(d) an affine plane with one point at infinity.

As usual, we break the proof of this theorem into a number of propositions in this section. Throughout, S is a finite {0,1}-semiaffine linear space. The first result is clear.

Proposition 4.2.1. If (p,L) is a non-incident point-line pair then $r_p \in \{k_L, k_L + 1\}$.

Proposition 4.2.2. If there exist two lines L and L' such that any point of S is on L or on L', then one of the following assertions holds:

(a) S is a near-pencil,
(b) S is the (3,3)-cross,
(c) S is the affine plane of order 2.

PROOF. Suppose first that L and L' intersect in a point p. If one of these lines has degree 2, then S is a near-pencil. So we may suppose that both L and L' have at least three points. It suffices to show that L and L' have precisely three points. But if either line has at least four points, we contradict the fact that S is {0,1}-semiaffine.

Now suppose that $L \cap L' = \phi$. If either line has more than two points, we again contradict the fact that S is $\{0,1\}$-semiaffine. □

Corollary 4.2.3. If there exist points p and p' such that every line is incident either with p or with p', then S is a near-pencil.

PROOF. Clearly any point q outside pp' has degree 2 and so all points lie on qp or qp'. Now apply Proposition 4.2.2. □

In view of Proposition 4.2.2 and its corollary, we suppose henceforth that given any two lines, there is a point on neither of them, and given any two points, there is a line through neither of them.

Let $n + 1$ be the maximal point degree of S. (Thus n is the *(Dembowski) order* of S, as mentioned in Section 1.4.)

In the propositions following, we determine the point and line degrees of S.

We shall let p_0 denote a point of degree $n + 1$ throughout.

Proposition 4.2.4. Any point has degree n or $n + 1$.

PROOF. Fix an arbitrary point $p \neq p_0$ and consider a line L on neither point. Applying Proposition 4.2.1 twice yields

$$k_L \in \{n, n+1\}$$
$$r_p \in \{n, n+1, n+2\}.$$

Since $r_p \leq n + 1$, the assertion follows. □

Proposition 4.2.5. Any line has degree n or $n + 1$.

PROOF. Since for any line L there is a point not on it, and since $r_{p_0} = n + 1$ is maximal, we have $k_L \leq n + 1$ for all L.

Suppose there is some line M with $k_M \leq n - 1$. By Propositions 4.2.1 and 4.2.4, any point not on M has degree n. Hence $k_M = n - 1$, and $p_0 \in M$. Moreover, for any line L, we can find a point on neither L nor M, and so $k_L = n - 1$ or n. In addition, if $p_0 \notin L$, then $k_L = n$.

Case 1. Suppose that M is the unique $(n - 1)$-line. Letting q be any point outside M, we get

$$\begin{aligned} n(n-1) = r_q(n-1) = v - 1 &= k_M - 1 + (r_{p_0} - 1)(n-1) \\ &= k_M - 1 + n(n-1), \end{aligned}$$

implying $k_M = 1$, a contradiction.

Case 2. There exist at least two $(n - 1)$-lines. Thus, p_0 is the intersection of all $(n - 1)$-lines, and any point other than p_0 has degree n. It follows that all lines on p_0 are $(n - 1)$-lines. Letting p denote any n-point, we have

$$n - 2 + (n - 1)^2 = n - 2 + (r_p - 1)(n - 1)$$
$$= v - 1 = r_{p_0}(n - 2) = (n + 1)(n - 2),$$

a contradiction. □

The next three propositions determine the structure of S completely.

Proposition 4.2.6. If there exists an n-point q, then S is an affine plane together with a point at infinity, q.

PROOF. By Propositions 4.2.1 and 4.2.5, any line not passing through q has degree n. By Propositions 1.6.1 and 4.2.5, there exists at least one $(n + 1)$-line. As any point outside an $(n + 1)$-line has degree $n + 1$, all $(n + 1)$-lines must pass through q.

We claim that there are at least two $(n + 1)$-lines. To see this, suppose L is a unique $(n + 1)$-line. Then any point $p \notin L$ has degree $n + 1$, and consequently

$$(n + 1)(n - 1) = r_p(n - 1) = v - 1$$
$$= k_L - 1 + (r_q - 1)(n - 1) = n + (n - 1)^2,$$

a contradiction.

Hence, all $(n + 1)$-lines pass through q, and any point different from q has degree $n + 1$.

It follows that the set of $(n + 1)$-lines is precisely the set of lines through q. Consequently, $v = 1 + r_q n = n^2 + 1$. $S - q$ is therefore an affine plane of order n. □

In view of Propositions 4.2.4 and 4.2.6, we may now assume that all points have degree $n + 1$.

Proposition 4.2.7. If both n- and (n + 1)-lines are present, then S is a punctured projective plane of order n.

PROOF. For any n-line M, let $\Pi(M)$ be the set of all lines missing or equal to M. Since all points have degree $n + 1$, there is precisely one line of $\Pi(M)$ through each point.

Consider an $(n + 1)$-line L. All lines of $\Pi(M)$ meet L. So $|\Pi(M)| = n + 1$.

As all lines meet any $(n + 1)$-line, $\Pi(M)$ consists entirely of n-lines. So $v = |\Pi(M)| \cdot n = n^2 + n$.

For any fixed point, let a denote the number of n-lines through it. Then

$$v - 1 = n^2 + n - 1 = a(n - 1) + (n + 1 - a)n = (n + 1)n - a,$$

implying $a = 1$. Therefore, $\Pi(M)$ is the set of all n-lines.

Extending S by a point at infinity corresponding to $\Pi(M)$ results in a projective plane of order n. □

Proposition 4.2.8. (a) If all lines are $(n + 1)$-lines, then S is a projective plane of order n.

(b) If all lines are n-lines, then S is an affine plane of order n.

PROOF. This follows immediately from the fact that each point has degree $n + 1$. □

4.3 $\{0,s,t\}$-Semiaffine linear spaces with non-constant point degree

This section gives a description of all $\{0,s,t\}$-semiaffine linear spaces for $0 < s < t$, $t \neq 2s$, in which not all points have the same degree. The next two sections will consider the case of constant point degree.

First of all, we introduce some trivial examples of $\{0,s,t\}$-semiaffine linear spaces.

Let S be a linear space and suppose that there are two lines L_1 and L_2 with respective degrees $s_1 + 1$ and $s_2 + 1$, such that any point is on one of these lines. It is easy to see that if L_1 and L_2 are disjoint, then S is $\{1,s_1,s_2\}$-affine. However, if L_1 and L_2 meet, then S is $\{0\}$-affine (a near-pencil), or $\{0,s_1 - 1,s_2 - 1\}$-affine.

Let s and t be non-negative integers with $s < t$. Define $S_{s,t}$ to be the unique linear space with $t + 3$ points and exactly one line of degree $t - s + 2$, while every other line has two points. Then $S_{s,t}$ is $\{s,t\}$-affine of order $t + 1$ (see Section 1.4 for the definition of *order*), and with point degrees $s + 2$ and $t + 2$.

Finally, we point out that for $s \geq 1$, finite $\{s\}$-affine linear spaces are precisely $2 - ((k + s)(k - 1) + 1,k,1)$ block designs. For $s = 0$, it suffices to include near-pencils in order to get a characterization.

The aim of the next series of propositions is to determine the line degrees of S, as summarized in Theorem 4.3.6 below.

We assume throughout, that S is a finite $\{0,s,t\}$-semiaffine linear space of order n with $0 < s < t$, $t \neq 2s$, such that S is neither the union of two lines, nor $S_{s,t}$.

Let p_0 denote a fixed $(n + 1)$-point.

Proposition 4.3.1. (a) Any line which is not incident with all $(n + 1)$-points has degree $n + 1$, $n + 1 - s$ or $n + 1 - t$.

(b) Any point has degree $n + 1$, $n + 1 - s$, $n + 1 - t$, $n + 1 - t + s$.

(c) Any line has degree $n + 1$, $n + 1 - s$, $n + 1 - t$, $n + 1 - t + s$, $n + 1 - 2t + s$, $n + 1 - 2s$, $n + 1 - 2t$, or $n + 1 - t - s$.

PROOF. (a) If p denotes an $(n + 1)$-point outside the line L, then $k_L \in \{r_p, r_p - s, r_p - t\} = \{n + 1, n + 1 - s, n + 1 - t\}$.

(b) Consider a point $p \neq p_0$. Since S is not the union of two lines, through a fixed point $q \notin pp_0$ there is a line $L \neq pp_0$. By (a), $k_L \in \{n + 1, n + 1 - s, n + 1 - t\}$. Moreover, $p \notin L$ implies $r_p \in \{k_L, k_L + s, k_L + t\}$. Thus $r_p \in \{n + 1, n + 1 - s, n + 1 - t, n + 1 - t + s\}$.

(c) Let L be an arbitrary line of S. Considering a point $p \notin L$ and applying (b) gives the result. □

Proposition 4.3.2. There is no line of degree $n + 1 - 2t + s$ or $n + 1 - 2t$.

PROOF. Assume, to the contrary, that there is a line L of one of the degrees in question. By Proposition 4.3.1(b), any point not on L has degree $k_L + t \neq n + 1$, since $t \neq 2s$, unless L has degree $n + 1 - 2t + s = n + 1 - 2s$ in which case $3s = 2t$, and each point not on L has degree k_L or $k_L + s$.

Suppose $3s \neq 2t$. Then $p_0 \in L$, and any line M which does not pass through p_0 has degree $n + 1$, $n + 1 - s$ or $n + 1 - t$ by Proposition 4.3.1(a), but also has degree $k_L + t$, k_L or $k_L + t - s$, since there exists a point not on $L \cup M$. However, this implies $k_M = n + 1 - t$.

Consider distinct lines X and Y on p_0 and different from L. Let $x \in X$, $y \in Y$, $x, y \neq p_0$. Then $(k_L + t - 1)(n - t) + k_{xp_0} = (k_L + t - 1)(n - t) + k_{yp_0}$ implies that all lines on p_0 have the same degree, say k. Clearly, $k = k_L + t$, k_L or $k_L + t - s$.

Now counting v using p_0 and a point not on L yields $k_L + n(k - 1)$

$= v = k + (k_L + t - 1)(n - t)$. This gives a contradiction unless $k = k_L + t - s$. In this case, consider a point $p \in L$, $p \neq p_0$. Then

$$v = k_L + (r_p - 1)(n - t) = k + (k_L + t - 1)(n - t)$$

implies $n - t | t - s$, and so $n - t \leq t - s$. It follows that $n + 1 - 2t < n + 1 - 2t + s \leq 1$ which is impossible.

We therefore turn to the case $3s = 2t$, $k_L = n + 1 - 2s$, and by Proposition 4.3.1(b), each point not on L has degree $k_L + s$ or $k_L + t$. By Proposition 4.3.1(a), $p_0 \in L$, and any line M not on p_0 has degree $n + 1$, $n + 1 - s$ or $n + 1 - t$. However, M also has degree k_L, $k_L + s$, $k_L + t$, $k_L + s - t$ or $k_L + t - s$, forcing $k_M = n + 1 - s$ or $n + 1 - t$.

Suppose there is a line M of degree $n + 1 - s$. Let $p \notin M \cup L$. Then $r_p = n + 1$, or $n + 1 - s$ because $p \notin M$, and $r_p = n + 1 - s$ or $n + 1 - 2s + t$ because $p \notin L$. So $r_p = n + 1 - s$. It follows that every line on p meets M. Consequently, all lines, except possibly L, meet M. However, there are s lines on p_0 which cannot meet M, and so we have a contradiction unless M is on p_0. We conclude that any $(n + 1 - s)$-line is on p_0, and any line not on p_0 has degree $n + 1 - t$. Moreover, all points except possibly points of $M \cup L$ have degree $n + 1 - s$.

Let $p \neq p_0$, $p \in M$. Using also a point of L different from p_0 and counting v in two different ways, we obtain

$$(n - s)(n - t) + n + 1 - 2s = (r_p - 1)(n - t) + n + 1 - s,$$

yielding $r_p = n + 1 - s - s/(n - t)$. But $r_p = k_L + s = n + 1 - s$ or $r_p = k_L + t = n + 1 - 2s + t$. So $0 < s/(n - t) = s - t < 0$.

We are now able to state that all lines not on p_0 have degree $n + 1 - t$, and that there are no lines of degree $n + 1 - s$ or of degree $n + 1$.

Let $p \notin L$. Then there is a line M not on p_0 and not on p. Thus M is an $(n + 1 - t)$-line. Then $r_p = n + 1 - \frac{s}{2}$, $n + 1 - s$ because $p \notin L$, and $r_p = n + 1$, $n + 1 - \frac{s}{2}$ or $n + 1 - \frac{3s}{2}$ because $p \notin M$. Hence $r_p = n + 1 - \frac{s}{2}$ for all $p \notin L$.

Let $q \in L$, $q \neq p_0$, and $p \notin M \cup L$. Then

$$(r_q - 1)(n - t) + n + 1 - 2s = v = \left(n - \frac{s}{2}\right)(n - t) + k_{pp_0}. \quad (1)$$

We therefore obtain

$$n - t | k_{pp_0} + \frac{s}{2} - 1 > 0,$$

and so $n + 1 - 2s \leq k_{pp_0}$. By Proposition 4.3.1(c), there are three possibilities: $k_{pp_0} = n + 1 - 2s, n + 1 - t, n + 1 - t + s$. If $k_{pp_0} = n + 1 - t$, the divisibility condition above yields $n \leq 2s$, a contradiction.

The three possible values for r_q are $n + 1 - t$, $n + 1$ and $n + 1 - t + s$. Substituting any of these into equation (1) with $k_{pp_0} = n + 1 - t + s$ leads to a contradiction. Substituting any of these with $k_{pp_0} = n + 1 - 2s$ leads only to $r_q = n + 1 - t + s = n + 1 - s/2$. Thus all points different from p_0 have degree $n + 1 - s/2$, and all lines on p_0 have degree $n + 1 - 2s$.

Finally, fixing any point $p \neq p_0$ and counting points in two ways, we obtain

$$n(n - 2s) = \left(n - \frac{s}{2}\right)(n - t),$$

implying either s or t equals 0, a contradiction. □

Proposition 4.3.3. There is no line of degree $n + 1 - t - s$.

PROOF. Assume such a line exists. Call it L. Using Proposition 4.3.1(b), any point not on L has degree $n + 1 - s$ or $n + 1 - t$. In particular, $p_0 \in L$.

Suppose there are distinct points p_s and p_t not on L and of degrees $n + 1 - s$ and $n + 1 - t$ respectively. Since clearly $r_{p_0} \geq 4$, there is a line $M \neq L$, $p_0 \in M$, $p_s, p_t \notin M$. Then $k_M = n + 1 - s, n + 1 - s - t$ or $n + 1 - 2s$, since $p_s \notin M$, and $k_M = n + 1 - t, n + 1 - 2t, n + 1 - t - s$, since $p_t \notin M$. Thus $k_M = n + 1 - t - s$, since $t \neq 2s$. It follows that at most one line through p_0 does not have degree $n + 1 - t - s$. We thus examine two cases. Suppose first of all that all lines through p_0 have degree $n + 1 - t - s$. Hence $v = 1 + (n + 1)(n - t - s)$. Let M be any line on p_s different from p_0p_s and p_sp_t. Then since $p_0 \notin M$, $k_M = n + 1, n + 1 - s$, or $n + 1 - t$, while as above, $p_t \notin M$ implies $k_M = n + 1 - t$ or $n + 1 - t - s$. So $k_M = n + 1 - t$. Let p_sp_t have degree k. Then $v = k + n - t - s + (n - 1 - s)(n - t)$. This, together with the above equation in v, yields $k = n + 1 - t - st$, which is impossible by Propositions 4.3.1(c) and 4.3.2.

We therefore suppose that there is a single line M on p_0 with degree k'

$\neq n + 1 - t - s$. Then $v = k' + n(n - t - s)$. If p_s can be chosen such that $p_s \notin L \cup M$, then arguing as above,

$$k' + n(n - t - s) = v = k + n - t - s + (n - 1 - s)(n - t),$$

implying $k' - k = s(t - 1)$. Then $p_s \notin M$ implies $k' \in \{n + 1 - s, n + 1 - 2s\}$. So $k = k' - s(t - 1) \in \{n + 1 - st, n + 1 - st - s\}$. Propositions 4.3.1(c) and 4.3.2 now imply $s = 1$. From $p_0 \notin p_s p_t$, it follows that $k = n + 1 - t$ and so $k' = n = n + 1 - s$. From the first part of the proof, we may assume $p_t \in M$. Thus, counting v in two ways using p_0 and p_t yields

$$k' + n(n - t - s) = k' + k - 1 + (n - 1 - t)(n - s)$$

or $n - t - 1 = n - t$, a contradiction.

We have now reached the conclusion that each point outside L has degree $n + 1 - s$ or each point outside L has degree $n + 1 - t$. Let this fixed degree be r. Then any line not through p_0 has degree r also. It also follows that all lines on p_0 different from L have fixed degree $k \in \{r, r - s, r - t\}$. Hence

$$k + (r - 1)^2 = v = k_L + n(k - 1),$$

or

$$k(n - 1) = r(r - 2) + s + t.$$

The only situation not resulting in a contradiction here is $r = n + 1 - t$, $k = r - s = n + 1 - t - s$. So *all* lines on p_0 have degree $n + 1 - t - s$, all points except p_0 have degree $n + 1 - t$ and any line not on p_0 has degree $n + 1 - t$. Fix a line M, $p_0 \notin M$. There is a line X on p_0 such that $X \cap M = \phi$. So through any point $p \neq p_0$ on X there is at least one line not meeting M, contradicting $r_p = n + 1 - t = k_M$. □

Proposition 4.3.4. There is no line of degree $n + 1 - t + s$.

PROOF. Let L be a line of degree $n + 1 - t + s$. It follows that any point not on L has degree $n + 1 - t + s$ or $n + 1 - t + 2s$. By Proposition 4.3.1(b), the latter case occurs only if $t = 3s$.

If points p_{t-s} and p_{t-2s} of both types occur in $S - L$, then any line not on p_0 or on p_{t-s} has degree $n + 1 - t$ (using the assumption $t \neq 2s$), and any line not on p_{t-s} or on p_{t-2s} has degree $n + 1 - t + s$ (again using $t \neq 2s$). Hence, any line goes through one of the three points p_0, p_{t-s}, p_{t-2s}, and any point different from these three on one of the lines p_0p_{t-s}, p_0p_{t-2s},

$p_{t-s}p_{t-2s}$ has degree 2. Since S is not the union of two lines, this cannot happen.

Thus, suppose that any point not on L has degree r. If $r = n + 1 - t + 2s$, then any line not on p_0 has degree $r = n + 1 - s$. If $r = n + 1 - t + s$, then any line not on p_0 has degree $n + 1 - t$. In either case, arguing as at the end of the proof of Proposition 4.3.3 leads to a contradiction. □

Proposition 4.3.5. There is no line of degree $n + 1 - 2s$.

PROOF. Let L be a line of degree $n + 1 - 2s$. Using Proposition 4.3.1(b), it follows that any point not on L has degree $n + 1 - s$, $n + 1 - 2s$ (implying $t = 3s$, since $t \neq 2s$), or $n + 1 - 2s + t$ (implying $2t = 3s$, since $t \neq 2s$).

Assume that there exist an $(n + 1 - s)$-point p_s and a point p of degree either $n + 1 - 2s$ or $n + 1 - 2s + t$, $p, p_s \notin L$. Then any line contains at least one of the points p_0, p, p_s, arguing as in the proof of Proposition 4.3.4. Consider first of all the possibility that any point lies on one of the lines p_0p, p_0p_s, pp_s. Then any other line has degree 2 or 3. In particular, any line on p_0, p or p_s has degree 2, except possibly for the lines p_0p, p_0p_s, pp_s themselves. Since S is not the union of two lines, each of p_0p, p_0p_s, pp_s has degree at least 3, implying that there exists a line on none of p_0, p, p_s, a contradiction. Thus, there is a point not on any of p_0p, p_0p_s, pp_s. It follows that each of these three lines has degree 2. In particular $2 \in \{n + 1, n + 1 - s, n + 1 - t\} \cap \{n + 1 - s, n + 1 - 2s\} = n + 1 - s$, since $t \neq 2s$. But then $n + 1 - 2s < 2$, a contradiction.

Now suppose that any point not on L has degree $n + 1 - s$. Since $t \neq 2s$, it follows that any line not on p_0 has degree $n + 1 - s$. Let M be a fixed line not on p_0, and M$'$ a line on p_0 missing M. Then there is a point of M' of degree $n + 1 - s$ which can therefore be on no line missing M – a contradiction.

We may finally suppose that each point not on L has the same fixed degree r where either $r = n + 1 - 2s$ or $r = n + 1 - 2s + t$. It follows, using $t = 3s$ or $2t = 3s$ as appropriate, that each line not on p_0 has degree $n + 1 - t$. Let k be the degree of a line on p_0 different from L. If $r = n + 1 - 2s$, then using Propositions 4.3.2, 4.3.3 and 4.3.4, we see that $k = n + 1 - 2s$ or $n + 1 - t$. If $r = n + 1 - 2s + t$, then the same propositions allow us to draw the same conclusions. Now counting points in two ways leads to $n + 1 - 2s + n(k - 1) = k + (r - 1)(n - t)$, while all possible choices for r and k lead to contradictions. □

We have to this point established the line degrees of S:

Theorem 4.3.6. Let S be a finite $\{0,s,t\}$-semiaffine linear space of order n with $0 < s < t$, $t \neq 2s$, such that S is neither the union of two lines nor $S_{s,t}$. Then any line of S has degree $n + 1$, $n + 1 - s$ or $n + 1 - t$.

We now attempt to determine the point degrees of S.

Theorem 4.3.7. Let S be a finite $\{0,s,t\}$-semiaffine linear space of order n with $0 < s < t$, $t \neq 2s$, such that S is neither the union of two lines nor $S_{s,t}$.

(a) If there exists a point p of degree $n + 1 - s$ or $n + 1 - t$, then $S - p$ is the pseudo-complement of an s-pencil or t-pencil, respectively, in a projective plane of order n.
(b) If S has a point of degree $n + 1 - t + s$, then there exists an $(n + 1)$-line L such that $S - L$ is a block design with $n - t$ points per line, admitting a parallelism with precisely $n + 1$ parallel classes. Moreover, $t - s$ is a divisor of s.

PROOF. (a) Proposition 4.3.1(b) and Theorem 4.3.6 imply that any point has degree $n + 1$, $n + 1 - s$, $n + 1 - t$ or $n + 1 - t + s$ and any line has degree $n + 1$, $n + 1 - s$ or $n + 1 - t$. Write $r_p = n + 1 - a$ where $a = s$ or t. Then, since $t \neq 2s$, any line not on p also has degree $n + 1 - a$, and it follows that each point has degree $n + 1$, $n + 1 - a$ or $n + 1 - t + s$, the last value occurring only if $a = t$.

Suppose there is a second point p' of degree $n + 1 - a$. Then all lines except possibly pp' have degree $n + 1 - a$. Letting k denote the degree of pp' yields, since p_0 cannot be on pp',

$$k + (n - a)^2 = v = 1 + (n + 1)(n - a)$$

or

$$k = 1 + (n - a)(a + 1),$$

which in turn implies $k = n + 1$ and $n = a + 1$, and so $n + 1 - a = 2$ and S is a near-pencil, contradicting our assumption.

Thus, p is unique. If there is an $(n + 1 - t + s)$-point q, then $a = t$, and any line different from pq has degree $n + 1 - t$. If k denotes the degree of pq, we obtain the contradiction

$$k + (n - t)(n - t) = v = k + (n - t + s)(n - t).$$

So all points except p have degree $n + 1$. Consequently, all lines on p have the same degree k. So

$$1 + (n + 1 - a)(k - 1) = k + n(n - a),$$

which implies $(n + 1 - k)(n - a) = 0$, and so $k = n + 1$. In this case, $S - p$ is the pseudo-complement of an a-pencil in a projective plane of order n.

(b) Let p be a point of degree $n + 1 - t + s$. Then, arguing as in (a), any line not on p has degree $n + 1 - t$, and any point has degree $n + 1 - t$, $n + 1 - t + s$ or $n + 1$. Because of (a), we may assume that no points of degree $n + 1 - t$ exist.

Suppose there is a second point p' of degree $n + 1 - t + s$. Then any line different from pp' has degree $n + 1 - t$. It follows that any point on pp' has degree $n + 1 - t + s$, while any point not on it has degree $n + 1$. Letting k be the degree of pp', we obtain

$$(n + 1)(n - t) = v - 1 = k - 1 + (n - t + s)(n - t),$$

or

$$k = 1 + (n - t)(t + 1 - s) > n + 1 - t.$$

Once again, using the argument invoked in (a), we see that $k = n + 1 - s$ or $n + 1$. If the former value is correct, then $n + 1 - s = 1 + (n - t)(t + 1 - s)$ implies $n = t + 1$, and so any line other than pp' has degree $n + 1 - t = 2$. Thus $S = S_{s,t}$, a contradiction. If the latter value is correct, then all lines meet pp'. Then $S - pp'$ is a block design, $2 - (v',n - t,1)$, admitting a parallelism with precisely $n + 1$ parallel classes, each parallel class corresponding to a point of pp'. Furthermore, $n + 1 = 1 + (n - t)(t + 1 - s)$ yields $(n - t)(t - s) = t$ and so $t - s|t$, from which it follows that $t - s|s$.

Finally, suppose that any point different from p has degree $n + 1$. Then each line on p has constant degree k, implying

$$(n + 1 - t + s)(k - 1) = v - 1 = k = 1 + n(n - t),$$

which contradicts $k \in \{n + 1,n + 1 - t,n + 1 - s\}$ □

The following important corollary to the above theorem will be needed in the next section.

Corollary 4.3.8. Let S be a {0,1,s}-semiaffine linear space of order n, with $s \geq 3$, and which is not {0,1}-semiaffine. Suppose, in addition, that not all points have the same degree. Then one of the following holds:

(a) $S = S_{1,s}$ or S is a $(3, s + 2)$-cross.
(b) S is a 2-$((k + s)(k - 1) + 1,k,1)$-design for some positive integer k.
(c) There is a point p such that $S - p$ is the pseudo-complement of an s-pencil in a projective plane of order n.

We state the next theorem without proof. The reader is encouraged to develop a proof using the methods of this section.

Theorem 4.3.9. Let S be a {1,2}-affine linear space of order n. Then any point has degree $n + 1$ and the line degrees are $n + 1$ and n except for the case $S = S_{1,2}$.

4.4 $\{s - 1,s\}$-Semiaffine linear spaces with constant point degree

In this section and the next, we consider the constant point degree case. Rather than making arguments easier, the homogeneity assumed in these sections seems to act as a barrier to a determination of the full structure of the space. We are not able to give results for the complete situation of Section 4.3.

In the present section, we restrict ourselves to $\{s - 1,s\}$-semiaffine linear spaces with constant point degree. Note that $s = 1$ is the Kuiper–Dembowski case, which was handled in Section 4.2. We therefore suppose $s \geq 2$.

Clearly, each line has either $n + 1 - s$ or $n + 2 - s$ points, and each point is on the same number of $(n + 1 - s)$- and of $(n + 2 - s)$-lines.

Let σ be the number of $(n + 1 - s)$-lines on any point, and let $b' = b_{n+1-s}$ be the total number of $(n + 1 - s)$-lines. We thus obtain the following equations.

$$v - 1 = (n + 1)(n + 1 - s) - \sigma \tag{2}$$

$$b' (n + 1 - s) = v\sigma = [(n + 1)(n + 1 - s) - \sigma + 1] \tag{3}$$

$$\begin{aligned}(b - b')(n + 2 - s) &= v(n + 1 - \sigma) \\ &= [(n + 1)(n + 1 - s) - \sigma + 1]\,(n + 1 - \sigma)\end{aligned} \tag{4}$$

Equations (3) and (4) imply the existence of integers x (non-negative) and y such that

$$(n+1-s)x = \sigma(\sigma-1) \tag{5}$$

$$(n+2-s)y = (\sigma+s-2)(\sigma+1-s). \tag{6}$$

Then (5) and (6) together give

$$\begin{aligned} y+(n+1-s)y = (n+2-s)y &= (\sigma-1+s-1)(\sigma+1-s) \\ &= (\sigma-1)\sigma+s-1-(s-1)^2 \\ &= (n+1-s)x-(s-1)(s-2). \end{aligned} \tag{7}$$

It follows from equation (7) that $n+1-s \mid y+(s-1)(s-2)$.

Proposition 4.4.1. We have $y+(s-1)(s-2) \geq 0$. Equality holds if and only if $s=2$ and S is an affine plane or a punctured affine plane.

PROOF. Assume $y+(s-1)(s-2) < 0$. Then equation (7) implies

$$(n+1-s)x-(s-1)(s-2) < -(n+2-s)(s-1)(s-2).$$

So

$$(n+1-s)(x+(s-1)(s-2)) < 0.$$

Since $n+1-s > 0$, we get

$$0 > x+(s-1)(s-2) \geq 0,$$

a contradiction.

Suppose, then, that $y+(s-1)(s-2) = 0$. From equation (7), we get $x = y \geq 0$; subsequently $y = 0 = x$ and $s = 2$. In view of equation (5) now, $\sigma = 0$ or 1. If $\sigma = 0$, then S is an affine plane of order n. If $\sigma = 1$, equations (2) and (3) imply $v = n^2-1$ and $b' = n+1$. Moreover, the $(n+1-s)$-lines partition the points. Adjoining a point at infinity corresponding to this partition yields an affine plane of order n. Thus, S is a punctured affine plane of order n. □

For the remainder of the section, we assume $y+(s-1)(s-2) > 0$.

Proposition 4.4.2. Either $n \leq s^2-1$, or σ satisfies

$$\sigma^2-\sigma-(s-1)(s-2)-(n+2-s)(n-(s-1)^2) = 0.$$

In the latter case, we get in particular: If $s=2$, then S is the pseudo-

complement of two lines in a projective plane of order n; if s = 3, then $\sigma = n - 2$ and S is the pseudo-complement of a triangle in a projective plane of order n; if $s \geq 4$, then $n \leq (s^4 - 6s^3 + 13s^2 - 8s - 1)/4$.

PROOF. Since $y + (s-1)(s-2) > 0$, we can use equation (7) to write

$$(n+1-s)\,z = y + (s-1)(s-2) \tag{8}$$

for some positive integer z. Suppose first of all that $z \geq 2$. Since $\sigma \leq n + 1$, equation (6) implies $(n + 2 - s)\,y \leq (\sigma + s - 2)(n + 1 + 1 - s)$, and hence $y \leq \sigma + s - 2$. Therefore,

$$\begin{aligned} 2(n+1-s) \leq y + (s-1)(s-2) &\leq \sigma + s - 2 + (s-1)(s-2) \\ &\leq n + 1 + s - 2 + (s-1)(s-2), \end{aligned}$$

from which we obtain $n \leq s^2 - 1$.

Now suppose $z = 1$, and so $y = n - (s-1)^2$. Substituting in equation (6) gives

$$(n+2-s)(n-(s-1)^2) = \sigma^2 - \sigma - (s-1)(s-2).$$

Solving this quadratic in σ we get as discriminant

$$\Delta = 1 + 4(n^2 - s^2n + sn + n + s^3 - 3s^2 + 2s).$$

If $s = 2$, this equation reduces to $\Delta = 1 + 4\,(n^2 - n) = (2n - 1)^2$. So $\sigma = (1 \pm (2n-1))/2$. The non-negative solution is $\sigma = n$. Using equations (2), (3) and (4) we obtain $v = n^2 - n$, $b' = n^2$, $b = n^2 + n + 1 - 2$, and so S is the pseudo-complement of two lines in a projective plane of order n.

If $s = 3$, $\Delta = (2n - 5)^2$, implying $\sigma = n - 2$. Consequently, by equations (2), (3) and (4), $v = (n-1)^2$, $b' = (n-1)^2$ and $b = (n-1)^2 + 3\,(n-1)$. So S is the pseudo-complement of a triangle in a projective plane of order n. Finally, if $s \geq 4$,

$$\Delta < (2n - s^2 + s + 1)^2. \tag{9}$$

If $2n - s^2 + s + 1 \leq 0$, then $n < s^2 - 1$. On the other hand, if $2n - s^2 + s + 1 > 0$, then equation (9) implies $\Delta \leq (2n - s^2 + s)^2$, which reduces to $4n \leq s^4 - 6s^3 + 13s^2 - 8s - 1$. □

The above results essentially summarize what is known about $\{s-1, s\}$-semiaffine linear spaces with constant point degree and $s \geq 2$. However, we are able to give more information in the specific cases outlined in the following proposition.

Proposition 4.4.3. (a) A $\{1,2\}$-semiaffine linear space of order 3 is $S_{1,2}$, K_5, or can be obtained from an affine plane of order 3 by removing nothing, a single point, or all points of a line along with the line.

(b) A $\{2,3\}$-semiaffine linear space of order n is the pseudo-complement of a triangle, a (hypothetical) block design 2-(46,6,1) or K_6.

PROOF. (a) Since any point is incident with at most four lines, any line has only two or three points.

From equation (6): $3y = \sigma(\sigma - 1)$. Hence $y \geq 0$. Proposition 4.4.1 handled the case $y = 0$, so we assume $y > 0$. If $z \geq 2$, then by equation (8),

$$4 = 2(n + 1 - s) \leq y \leq \sigma \leq n + 1 = 4.$$

So $\sigma = n + 1$ and consequently all lines are 2-lines. Therefore, S is K_5. If $z = 1$, then by Proposition 4.4.2, S is the complement of a line in an affine plane of order 3.

(b) By Proposition 4.4.2, we have $n \leq 3^2 - 1 = 8$. In case $n = 8$, equation (8) implies $\sigma = n + 1$; so S is a block design in which any line has $n + 1 - s = 6$ points. Hence $v = 1 + (n + 1)(n - s) = 46$.

In any case, equations (5) and (6) read

$$(n-2)\,x = \sigma\,(\sigma - 1),\ (n - 1)y = (\sigma + 1)(\sigma - 2).$$

If $4 \leq n \leq 7$, we have only the following possibilities: $n = 4$ and $\sigma = 2$ or 5; $n = 5$ and $\sigma = 3$ or 6; $n = 6$ and $\sigma = 4$; $n = 7$ and $\sigma = 5$. If $n = 4$ and $\sigma = 5$, then any line is a 2-line and $S = K_6$. If $n = 5$ and $\sigma = 6$, S is an $S(2, 3, 13)$. In all other cases, S is the pseudo-complement of a triangle in a projective plane of order n. □

Summarizing the results of this section, and applying the characterizations of the pseudo-complement of two lines and of a triangle developed in the last chapter, we obtain the following result.

Theorem 4.4.4. Let S be an $\{s - 1,s\}$-semiaffine linear space of order n other than $S_{s-1,s}$, where $s \geq 2$.

(a) (Oehler 1975.) If $s = 2$, then either S can be obtained from an affine plane of order n by removing nothing, a single point, or all points of a line along with the line, or S is the Nwankpa–Shrikhande plane, K_5, K_6 or a 2 − (46,6,1) design.

(b) (Beutelspacher and Meinhardt 1984.) If $s = 3$ and $n \geq 6$, then S is the complement of a triangle in a projective plane of order n.

(c) (Beutelspacher and Meinhardt 1984.) For any $s \geq 4$, there exist at most finitely many $\{s - 1, s\}$-semiaffine linear spaces.

In the next section we discuss $\{0,1,s\}$-semiaffine linear spaces, $s \geq 2$, in which each point has degree $n + 1$. We shall discover that the linear spaces appearing in this guise are of greater variety than the ones we encountered in the present section.

4.5 $\{0,1,s\}$-Semiaffine linear spaces with constant point degree

We suppose $s \geq 2$, and that the degree of each point is $n + 1$. Thus any line has degree $n + 1$, n or $n + 1 - s$.

The results of this section are due to A. Beutelspacher and A. Kersten (1984).

Theorem 4.5.1. Let S be a $\{0,s\}$-affine linear space. Then S is a pseudo-complement of a maximal s-arc in a projective plane of order n.

PROOF. In this case, any line has degree $n + 1$ or $n + 1 - s$. Letting σ be the number of $(n + 1 - s)$-lines on a fixed point, we see that

$$v - 1 = \sigma(n - s) + (n + 1 - \sigma)n = (n + 1)n - \sigma s,$$

implying that σ is independent of the choice of point.

Since all lines meet an $(n + 1)$-line, we have $b = 1 + (n + 1)n$. Moreover, $b_{n+1-s} = (n + 1)\sigma$. Counting incident pairs (p,L) where L is an $(n + 1 - s)$-line, we also get $b_{n+1-s} = v\sigma/(n + 1 - s)$. It follows that $\sigma = [s(n + 1) - n]/s$, and $v = (n + 1)(n + 1 - s)$. Hence S is the pseudo-complement of a maximal s-arc in a projective plane of order n. □

From now on, we assume that there is at least one n-line in S.

For any n-line L, define

$$\Pi(L) = \{M \in \mathcal{L} | M = L \text{ or } M \cap L = \phi\}.$$

The following proposition follows immediately from the fact that any point has degree $n + 1$.

Proposition 4.5.2. Let L be an n-line.

(a) $\Pi(L)$ is a parallel class.

(b) *The lines of $\Pi(L)$ are of degree n or $n + 1 - s$.*
(c) *$\Pi(L)$ induces an equivalence relation on the lines of degree n.*

At this point, and for the remainder of the section, we introduce an additional assumption:

$$b \leq n^2 + n + 1.$$

We note that if $(n + 1)$-lines exist, then $b = n^2 + n + 1$.

Proposition 4.5.3. Either S is the pseudo-complement of an s-pencil in a projective plane of order n, or $n^2 + n \leq b$.

PROOF. As noted above, if $(n + 1)$-lines exist, we are done. So suppose S has only n- and $(n + 1 - s)$-lines. It follows that each point is on the same number of n- and of $(n + 1 - s)$-lines.

Suppose first of all that each point is on a unique n-line. Then each $(n + 1 - s)$-line meets each n-line, as otherwise it is possible to find a point on $n + 2$ lines. Thus $b_n = n + 1 - s$. Hence

$$b = b_n + n^2 = n^2 + n + 1 - s,$$

and

$$v = (n + 1 - s)n.$$

Therefore S is the pseudo-complement of an s-pencil in a projective plane of order n.

Suppose then that there exist intersecting n-lines M and M'. Since through any point of M' there is exactly one line parallel to M, the number of lines meeting either M or M' is $n^2 + n$. Thus $b \geq n^2 + n$. □

Theorem 4.5.4. Let S be a $\{1,s\}$-affine linear space, $s \geq 2$, satisfying equation (10). Then one of the following assertions holds:

(a) *S is the pseudo-complement of an s-pencil in a projective plane of order n*
(b) *S can be obtained from the pseudo-complement of a maximal $(s - 1)$-arc in a projective plane of order n by removing an $(n + 1)$-line.*
(c) *S is the complement of a set of points X in a projective plane of order n such that each line of the plane meets X in 1 or s points.*

PROOF. If $b < n^2 + n$, then by Proposition 4.5.3, S is the pseudo-complement of an s-pencil. Therefore, suppose $b \geq n^2 + n$.

If $b = n^2 + n$, it follows that any parallel class Π as defined above has precisely n elements, and distinct parallel classes are disjoint. So each n-line determines a unique parallel class.

Denote by m the number of n-lines in a particular parallel class. Then $v = mn + (n - m)(n + 1 - s) = n(n + 1 - s) + m(s - 1)$. So m is independent of the parallel class. Again using σ for the number of $(n + 1 - s)$-lines on a fixed point, we obtain $v = n^2 - \sigma(s - 1)$ so that $m = n - \sigma$.

If $\sigma = 0$, then any line has degree n and S has n^2 points, in which case S is an affine plane. But an affine plane is not $\{1,s\}$-affine.

Thus $\sigma \geq 1$. Consequently, each parallel class has at least one $(n + 1 - s)$-line.

Suppose some $(n + 1 - s)$-line L is in no parallel class. So all n-lines meet L. Thus $b_n = (n + 1 - s)(n + 1 - \sigma)$. But this is also the number of n-lines meeting an $(n + 1 - s)$-line of any parallel class, and since each parallel class contains at least one n-line, this is a contradiction.

Therefore, the set of all parallel classes partitions the line set of S into $n + 1$ classes. We now adjoin $n + 1$ new points corresponding to these $n + 1$ parallel classes, and one new line L_∞ in such a way that L_∞ is incident with precisely the new points. The new structure S^* is a linear space with $v^* = n^2 - \sigma(s - 1) + n + 1$ points and $b^* = n^2 + n + 1$ lines, in which each point has degree $n + 1$ and each line, degree $n + 1$ or $n - s$. Finally, from

$$\frac{v(n + 1 - \sigma)}{n} = b_n = m + n(n - \sigma) = (n + 1)(n - \sigma),$$

we obtain $\sigma = n + 1 - n/(s - 1)$, and so $v^* = (n + 1)(n + 1 - (s - 1))$. Hence, S^* is the pseudo-complement of a maximal $(s - 1)$-arc in a projective plane of order n.

We now turn to the case $b = n^2 + n + 1$. In this case, any parallel class consists of precisely $n + 1$ lines, and two distinct parallel classes have precisely one common line. Letting m be the number of n-lines in a parallel class, we obtain $v = mn + (n + 1 - m)(n + 1 - s) = n^2 + n - (n - m + 1)(s - 1)$. Also, $v = n^2 - \sigma(s - 1)$, and so,

$$m(s - 1) = (n - \sigma + 1)(s - 1) - n.$$

Fix an $(n + 1 - s)$-line L. There are $b_n - (n + 1 - s)(n + 1 - \sigma) > 0$ n-lines missing L. For each such line M, $L \in \Pi(M)$. On the other hand, the m n-lines of $\Pi(M)$ all miss L, and determine $\Pi(M)$ uniquely. So the number of parallel classes on L is

$$\frac{b_n - (n+1-s)(n+1-\sigma)}{m} = \frac{m + n(n-\sigma) - (n+1-s)(n+1-\sigma)}{m}$$

$$= 1 + \frac{(n-\sigma)(s-1) - n - 1 + s}{m}$$

$$= 1 + \frac{m(s-1)}{m} = s.$$

We adjoin $n + 1$ new points corresponding to the $n + 1$ parallel classes to obtain a linear space S^* with constant point degree and constant line degree $n + 1$. So S^* is a projective plane of order n. □

Finally in this section, we characterize all 'proper' $\{0,1,s\}$-semiaffine linear spaces.

Theorem 4.5.5. Let S be a $\{0,1,s\}$-affine linear space of order n. Then S is the complement of a set of points X in a projective plane of order n, such that each line of the plane meets X in 0, 1 or s points.

PROOF. Since there exists an $(n + 1)$-line, we have $b = n^2 + n + 1$. Therefore any parallel class has precisely $n + 1$ lines, and two distinct parallel classes have precisely one common line.

For a fixed point p, let λ_p and μ_p be the number of $(n + 1)$- and of n-lines respectively on p. Clearly,

$$v = 1 + \lambda_p s + \mu_p(s-1) + (n+1)(n-s).$$

Let m be the number of n-lines in some fixed parallel class. Since no parallel class contains an $(n + 1)$-line, we have

$$v = m(s-1) + (n+1)(n+1-s).$$

It follows that m is independent of the parallel classes, and that

$$\mu_p = m + (n - \lambda_p s)/(s-1)$$

for any point p.

Fix an n-line L. Then, since all $(n + 1)$-lines meet L, $b_{n+1} = \Sigma_{p \in L} \lambda_p$. On the other hand, the only n-lines missing L are those in $\Pi(L)$. So

$$\begin{aligned} b_n &= m + \sum_{p \in L} (\mu_p - 1) \\ &= m + mn + \frac{n^2}{s-1} - n - \frac{s}{s-1} \sum_{p \in L} \lambda_p \\ &= m + n(m-1) + \frac{n^2}{s-1} - \frac{s}{s-1} b_{n+1}. \end{aligned}$$

Let M be a fixed $(n + 1 - s)$-line. The number of n-lines missing M is $b_n - \Sigma_{p \in M} \mu_p$. For any one of these lines, say L, M is contained in $\Pi(L)$. Thus, M is contained in precisely $(b_n - \Sigma_{p \in M} \mu_p)/m$ parallel classes. Using the expression above for μ_p, this becomes $(b_n - m(n + 1 - s) - n(n + 1 - s)/(s - 1) + sb_{n+1}/(s - 1))/m$. Substituting in here the above expression for b_n, we find that any $(n + 1 - s)$-line is contained in exactly s parallel classes.

If we now adjoin new points as in the proof of Theorem 4.5.4, we obtain a projective plane of order n in which the set of new points is precisely as described. □

4.6 {0,1,2}-Semiaffine linear spaces

In this, the final section of the chapter, we characterize all {0,1,2}-semiaffine linear spaces. Throughout, S denotes a {0,1,2}-semiaffine linear space of order n, and p_0 a fixed $(n + 1)$-point. In view of the results already obtained, we may suppose that S is neither {0,1}- nor {1,2}-semiaffine. Consequently, $n \geq 3$. For most of the remainder of this section, we assume $n \geq 3$. However, for Proposition 4.6.1(b), we assume $n \geq 6$. The remaining cases are left as an exercise. We may also assume throughout that S is not the union of two lines.

Proposition 4.6.1. (a) Any point of S has degree $n + 1$, n or $n - 1$, and any line of S has degree $n + 1$, n, $n - 1$ or $n - 2$.

(b) If $n \geq 6$, then S has a line of degree $n - 2$.

PROOF. (a) Arguing as in the proof of Proposition 4.3.1, we find that any point has degree $n + 1$, n or $n - 1$ and any line has degree $n + 1$, n, $n - 1$, $n - 2$ or $n - 3$.

Suppose there exists an $(n - 3)$-line L. Then any point not on L has degree $n - 1$. Therefore, $p_0 \in L$, and any line not on p_0 has degree in $\{n + 1, n, n - 1\} \cap \{n - 1, n - 2, n - 3\} = \{n - 1\}$. Then on p_0 there are two lines missing a fixed $(n - 1)$-line M, $p_0 \notin M$. On the other hand, letting p be a point on either of these lines, $p \neq p_0$, *no* line on p misses M, a contradiction.

(b) Let L be an $(n - 2)$-line. Then any point not on L has degree $n - 1$ or n, and so there are no $(n + 1)$-lines. If there exists an n-line M not on p_0, then any point not on $L \cup M$ has degree n and any n-line not on p_0 misses L. It follows that all points of L are $(n + 1)$-points and all n-lines miss L. Counting v, at p_0 and at $p \notin L \cup M$ yields the fact that there are $n - 3$ n-lines on p. Since $n \geq 6$, we contradict the fact that no n-line meets L. Thus any line not on p_0 has degree $n - 1$. Consequently, counting v, all points of L except p_0 have the same degree r.

Suppose $r = n + 1$. Thus all lines but L have degree $n - 1$. Then counting v in two ways using p_0 and a point $q \notin L$, we obtain

$$n - 2 + n(n - 2) = v = 1 + r_q(n - 2).$$

This implies $n - 2 | 1$. So $n = 3$ and $n - 2 = k_L = 1$, a contradiction.

Suppose $r = n$. Assume that for some line M, $M \cap L = \phi$. Then $k_M = n - 1$ and $r_p = n - 1$ for all $p \in M$. Suppose a third line N misses both L and M. Let $p \in N$. Then $r_p = n$ and p is on a second line N' missing L but meeting M. However, the point $q = N' \cap M$ has degree $n - 1$, and so N' meets L, a contradiction. Thus no line misses both L and M. On the other hand, any line meeting either L or M and not on p_0 meets both L and M. Thus

$$(n - 3)(n - 1) + n + 1 = b = (n - 1)(n - 2) + 1 + 2,$$

yielding $4 = 5$, a contradiction.

We conclude that all lines meet L. It follows that any point not on L has degree $n - 2$, a contradiction.

Finally, suppose $r = n - 1$. Then any line on p_0 has degree $n - 1$ or $n - 2$. If there exist both an $(n - 1)$-line L_1 and an $(n - 2)$-line L_2, $L_2 \neq L$, on p_0, letting $p \in L_1$ and $q \in L_2$, $p, q \neq p_0$, yields

$$n - 1 + (r_p - 1)(n - 2) = v = n - 2 + (r_q - 1)(n - 2) \qquad \text{or}$$

$$1 = (r_q - r_p)(n - 2).$$

Thus $n - 2 | 1$, a contradiction. So all lines different from L on p_0 have the same degree k. Then

$$n - 2 + n(k - 1) = v = n - 2 + (r - 1)(n - 2) = n - 2 + (n - 2)^2.$$

So $k - 1 = n - 4 + 4/n$. Consequently, $n = 4$, a contradiction. □

We assume henceforth that all lines have degree $n + 1$, n or $n - 1$ and $n \geq 3$.

Theorem 4.6.2. If S has an $(n - 1)$-point, then either S can be obtained from a projective plane of order n by removing all points on two lines but the point of intersection, along with the lines themselves, or S is the extended Nwankpa–Shrikhande plane.

PROOF. Clearly any line not on an $(n - 1)$-point p has degree $n - 1$. Since S is not a cross, we have $n - 1 = r_p \geq 3$. Also, since any line not on p meets each line on p, all lines on p have the same degree k, say. Therefore all points different from p have degree k. It follows that $k = n + 1$. Thus

$$v = 1 + (n - 1)n = n^2 - n + 1$$

and

$$b = r_p + n^2 = n^2 + n + 1 - 2.$$

By Proposition 3.3.3, S is as described. □

We assume from henceforth that all points have degree $n + 1$ or n. In view of the previous section, we may also assume that n-points exist.

Theorem 4.6.3. If S has exactly one line L of degree $n + 1$, then either there exists a projective plane P of order n, two lines L_1 and L_2 and two points p_1 and p_2 with $p_i \in L_i$, $p_i \neq L_1 \cap L_2$, such that $S = P - ((L_1 - p_1) \cup (L_2 - p_2))$, or S is K_6 extended by a line at infinity.

PROOF. Clearly, any point not on L has degree $n + 1$, all lines meet L and the number μ of n-lines through a point not on L is constant. Thus we have $v = n^2 - n + 1 + \mu$. But L contains an n-point, and so $v \leq n^2 - n + 2$. It follows that

$$v = n^2 - n - 1 + \mu, \qquad 0 \leq \mu \leq 3.$$

Note also that all n-points of L are on the same fixed number r of n-lines.

If every point of L has degree n, then $v = n^2 - 2n + 3 + r$. Comparison with the above equation for v yields $r = n - 4 + \mu$. Furthermore, $b_n = (n + 1)r$, and $b_n(n - 1) = (v - (n + 1))\mu$. So

$$(n - 4 + \mu)(n^2 - 1) = (n^2 - 2n - 2 + \mu)\mu.$$

For $\mu = 1$, 2 and 3, this last equation leads directly to a contradiction. The case $\mu = 0$ yields $n = 4$, $r = 0$ and $v = 11$. Consequently, S is K_6 with an additional line at infinity.

Suppose, then, that L contains an $(n + 1)$-point p. Denoting by r' the number of n-lines through p, we see that $v = n^2 - n + 1 + r'$. This, along with the above equation for v in terms of μ, yields $r' = \mu - 2$, $\mu = 2$ or 3.

If $\mu = 2$ and $r' = 0$, then $r = n - 2$. Letting β be the number of n-points on L, we obtain $b_n = \beta r = \beta(n - 2)$, while $b_n(n - 1) = (v - (n + 1))\, 2 = 2\,(n^2 - 2n)$, in which case $\beta = 2n/(n - 1)$, a contradiction.

If $\mu = 3$ and $r' = 1$, then $r = n - 1$. In this case, with β defined as above,

$$b_n = \beta r + (n + 1 - \beta)r' = \beta(n - 2) + n + 1.$$

Hence

$$(\beta(n - 2) + n + 1)(n - 1) = b_n(n - 1) = (v - (n + 1))3 = (n - 1)^2 3,$$

so $\beta = 2$.

Thus S, without the two points of degree n, is the pseudo-complement of two lines in a projective plane of order n. The result follows from Proposition 3.3.3. □

Theorem 4.6.4. If S has at least two $(n + 1)$-lines, then S can be obtained from a projective plane P of order n by removing a set q of points, where either q is the set of all points but one of a line L along with one point not on L, or q is the set of all points but one, p, of a line L, along with two points not on L, collinear with but distinct from p.

PROOF. Let L_1 and L_2 be distinct $(n + 1)$-lines, and $p = L_1 \cap L_2$. Then p is the unique n-point, all other points having degree $n + 1$. We distinguish three cases.

Case 1. There is an n-line N on p.

In this case, the set $\Pi(N)$ introduced in Section 4.5 is a parallel class. It is not difficult to see that all lines except N of $\Pi(N)$ have degree $n - 1$. Moreover, $|\Pi(N)| = n + 1$. Thus

$$v = n + (|\Pi(N)| - 1)(n - 1) = n^2.$$

It follows, that, except for N, all lines on p are $(n+1)$-lines. Since all $(n-1)$-lines meet the $n-1$ $(n+1)$-lines on p, all $(n-1)$-lines miss N. Thus the lines of $\Pi(N) - N$ are precisely the $(n-1)$-lines. Adding a point at infinity corresponding to $\Pi(N)$ and deleting p, we obtain a linear space S^* with constant line degree n and constant point degree $n+1$. Thus S^* is an affine plane of order n.

Case 2. All lines on p have degree $n+1$.
In this case, any line not on p is an n-line. Consequently, S is $\{0,1\}$-semiaffine, a contradiction.

Case 3. There is an $(n-1)$-line M on p but no n-line.
Arguing as in case 1, we have that any point not on M is on two lines missing M. Any such line is an $(n-1)$-line, and there are precisely $2n$ of these. Thus $(v-(n-1))2 = 2n(n-1)$, or $v = n^2 - 1$. Consequently, M is the unique $(n-1)$-line on p, all other lines on p being $(n+1)$-lines. It follows that $b_{n+1} = n-1$, $b_n = n(n-2)$, $b_{n-1} = 2n+1$. Therefore S is the complement of a punctured line along with two additional points, as described in the theorem. □

We may now assume that all lines have degree n or $n-1$.

Theorem 4.6.5. Suppose that the lines of S have degree n or n − 1 and the points degree n + 1 or n. Then one of the following holds:

(a) There exists a projective plane P, distinct lines L_1 and L_2 of P and a point p on L_1 or L_2, $p \neq L_1 \cap L_2$, such that $S = P - ((L_1 \cup L_2) - p)$.

(b) S is the Fano quasi-plane defined by $v = 7$, $b = 9$, $v_3 = 4$, $b_3 = 6$, $v_4 = 3$, $b_2 = 3$. (See Figure 6.1.1.)

(c) S is the punctured Fano plane or the Fano quasi-plane less a 3-point. (See Section 6.1.)

PROOF. Denote by μ and μ' the number of n-lines through a fixed $(n+1)$-point and n-point respectively. In view of Section 4.5, we may assume that points of both degrees exist. Thus

$$\mu + (n+1)(n-2) = v - 1 = \mu' + n(n-2),$$

or

$$\mu' - \mu = n - 2.$$

Since $0 \leq \mu$ and $\mu' \leq n$, we may restrict ourselves to the following cases.

1 $\mu' = n$, $\mu = 2$, $v = n^2 - n + 1$.
2 $\mu' = n - 1$, $\mu = 1$, $v = n^2 - n$.
3 $\mu' = n - 2$, $\mu = 0$, $v = n^2 - n - 1$.

Case 1. From the above we can see that no $(n-1)$-line contains an n-point.

Consider the two n-lines on p_0. Then any n-point lies on one of these lines. Consider a point p on one of the $(n-1)$-lines on p_0. Then p has degree $n + 1$, and all n-points lie on the two n-lines through p. Thus there are at most four n-points.

On the other hand,

$$\begin{aligned} b_n n &= v_{n+1}\mu + v_n\mu' \\ &= (v_{n+1} + v_n)\mu + v_n(\mu' - \mu) = v\mu + v_n(\mu' - \mu) \\ &= (n^2 - n + 1)2 + v_n(n - 2). \end{aligned}$$

Hence $n|2(v_n - 1)$.

Suppose $v_n \neq 1$. Then $v_n \leq 4$ implies $(v_n,n,v,b_n) = (4,3,7,6)$, $(4,6,31,13)$ or $(3,4,13,8)$. The first of these yields the Fano quasi-plane; while in the second and third cases, b_n is smaller than the number of n-lines through the n-points, a contradiction.

Thus $v_n = 1$. Let p be the unique n-point. Since $b_n = 2n - 1$, $S' = S - p$ is a linear space with $v' = n^2 - n$ points, all of which have degree $n + 1$. Furthermore, S' has $n - 1$ lines of degree n and n^2 lines of degree $n - 1$. So S' is the pseudo-complement of two lines in a projective plane of order n. It follows (Proposition 3.3.3) that S can be obtained from a projective plane of order n by removing a set $(L_1 \cup L_2) - p$ of points, where p is incident with precisely one of the lines L_1, L_2.

Case 2. In this situation, distinct n-lines meet in an n-point. Let L_1 and L_2 be distinct n-lines. If they are disjoint, then all points of $L_1 \cup L_2$ have degree $n + 1$. Thus, no n-line meets L_1 or L_2. But now, through any n-point, there are $n - 1$ lines of degree n all missing L_1, which is a contradiction. Thus any two n-lines meet.

Fix an $(n-1)$-line N. If two n-lines miss N, since they necessarily intersect, their point of intersection would have degree $n + 1$, a contradiction. So any $(n-1)$-line misses at most one n-line.

If the $(n-1)$-line N contains only $(n+1)$-points, the above paragraph and $\mu = 1$ imply $n - 1 \leq b_n \leq n$.

If there are at least two n-points in S then $b_n \geq 2\mu' - 1 = 2n - 3$, and so $n \leq 3$. Thus $n = 3$, $v = 6$, $\mu' = 2$, $\mu = 1$ and S is the punctured Fano plane.

If there is a unique n-point in S, then

$$b_n n = (v - 1) + 1 \cdot \mu' = n^2 - n - 1 + (n - 1),$$

which implies $n|2$, a contradiction.

Again, fix an $(n - 1)$-line N. By the above, N has at least one n-point p. If an n-line L misses N, we obtain a contradiction to degree p equals n. Thus *all* lines meet an n-line. It follows that all points not on an n-line have degree n. Since we may assume there are at least two n-points, any point misses some n-line, or is the intersection of two or more n-lines, and is therefore an n-point which, by a counting argument on v, leads to a contradiction.

Case 3. Counting incident point-line pairs (p,L) where L is an $(n - 1)$-line, we get in this case

$$\begin{aligned} b_{n-1}(n - 1) &= v_{n+1}(n + 1) + v_n^2 \\ &= 2(n^2 - n - 1) + v_{n+1}(n - 1). \end{aligned}$$

Consequently $n - 1|2$, and so $n = 3$. Thus $v = 5$ and $S = S_{1,2}$, a contradiction. □

Theorem 4.6.6. (Hauptmann 1980; Lo Re and Olanda 1984.) Let S be a {0,1,2}-semiaffine linear space of order $n \geq 6$, which is neither {0,1}- nor {1,2}-semiaffine. Then one of the following assertions is true.

(a) S is embeddable in a projective plane of order n as the complement of one of the following sets: (i) a set of points no three of which are collinear; (ii) the set of points of two lines, except for the point of intersection; (iii) the set of points of $(L_1 \cup L_2) - \{p_1,p_2\}$ where $p_1 \in L_1 - L_2$ and $p_2 \in L_2 - L_1$; (iv) the set of all points but one of a line L together with a point outside L; (v) the set of all points but one, p, of a line L together with two points not on L but collinear with p; (vi) the set of points of $(L_1 \cup L_2) - p$ where p is a point of L_1 or L_2, but not the point of intersection.

In addition, if $n \geq 3$ and there is no line of degree $n = 2$:

(b) S is the extended Nwankpa–Shrikhande plane.
(c) S is K_6 along with a line at infinity.
(d) S is the Fano quasi-plane, perhaps with a 3-point deleted.

4.7 Exercises

1. Show that any finite {1,2}-semiaffine linear space other than $S_{1,2}$ has constant point degree.

2. Is there a possible parameter set (v,b,n,s,σ) of a $\{1,s\}$-affine linear space with constant point degree $n + 1$ in which $b > n^2 + n + 1$?

   A $(v,3,3)$ *partial plane* is an incidence structure of v points and b lines (sets of points) such that any line has three points, any point is on three lines, and any two lines meet in at most one point. (These structures are also known as *v_3b_3-configurations*; see F. W. Levi 1929, for example.)
3. Show that there are unique (7,3,3)- and (8,3,3)-partial planes, while there are exactly two (9,3,3) partial planes.
4. Show that if S is a $\{2,3\}$-semiaffine linear space of order n with non-constant point degree, then $n = 4$ and S has three 5-points and three 4-points.
5. Characterize all $\{0,1,2\}$-linear spaces with $n = 3$, 4 or 5.
6. Determine all possible parameter sets for finite $\{4,5\}$-semiaffine linear spaces.

4.8 Research problems

1. Determine all $\{1,5\}$-semiaffine linear spaces of order n with $b > n^2 + n + 1$.
2. Characterize all $\{s_1,s_2,s_3\}$-semiaffine linear spaces, $0 < s_1 < s_2 < s_3$, with non-constant point degree.
3. Classify all $\{2,s\}$-semiaffine linear spaces.
4. Determine all $\{1,2,s\}$- and $\{2,3,4\}$-semiaffine linear spaces. (First convince yourself that interesting examples exist.)

4.9 References

Beutelspacher, A. and Kersten, A. (1984), Finite semiaffine linear spaces. *Arch. Math. 44*, 557–568.

Beutelspacher, A. and Meinhardt, J. (1984), On finite *h*-semiaffine planes. *Europ. J. Comb. 5*, 113–122.

Dembowski, P. (1962), Semiaffine Ebenen. *Arch. Math. 13*, 120–131.

Hauptmann, W. (1980), Endliche [0,2]-Ebenen. *Geom. Ded. 9*, 77–86.

Levi, F. W. (1929), *Geometrische Konfigurationen*. Hirzel, Leipzig.

Lo Re, M. and Olanda, D. (1986), On [0,2]-semiaffine planes. *Simon Stevin 60*, 157–182.

Oehler, M. (1975), Endliche biaffine Inzidenzebenen. *Geom. Ded. 4*, 419–436.

Pickert, G. (1955), *Projektive Ebenen*. Springer-Verlag, Berlin, Göttingen, Heidelberg.

5

Semiaffine linear spaces with large order

5.1 Introduction

In this chapter we consider semiaffine linear spaces of a much more general nature than in the previous chapter. Also, our methods are quite different. We take a somewhat graph-theoretic approach, using terminology and some basic notions from that area. For us, the vertices of the graph are the lines of the linear space.

Our principle result (Theorem 5.3.3) can be interpreted as follows. If the order n of the linear space S is large enough with respect to the difference between the maximum point degree and the minimum line degree in S, then S embeds in a projective plane of order n. This chapter is based on Beutelspacher and Metsch (1986, 1987).

Our first result is a very general one. The proof is broken into several stages by using propositions. For lines L and H, we denote by $m(L,H)$ the number of lines missing both L and H.

Theorem 5.1.1. Let $S = (p,\mathcal{L})$ be a linear space. Suppose there are a line H and integers a, c, d, e, n and x with the following properties.

(1) The degree of H is $n + 1 - d > 0$.

(2) The number of lines missing H is $nd + x > 0$.

(3) For every line L missing H we have $n - 1 + a \leq m(L, H) \leq n - 1 + c$.

(4) For any two intersecting lines L_1, L_2 *missing H, we have* $m(L_1,L_2) \leq e + 1$.

(5) (i) $2n > (d + 1)(de - 2a) + 2x$, *or*

(ii) there do not exist $d + 1$ *mutually intersecting lines missing H.*

(6) $n > (2d - 1)c + e - 2x$.
(7) $e \geq 0$.

Then the following hold:

(a) If $M_1, M_2, \ldots, M_s$ *are maximal sets of mutually disjoint lines with* $H \in M_i$ *and* $|M_i| \geq n - (d - 1)c + x + 1$, *then* $s = d$ *and every line disjoint from H is contained in exactly one of the sets* M_i.
(b) If every point not on H has degree $n + 1$, *then the sets* M_i *are parallel classes.*

The propositions involved in the proof of Theorem 5.1.1 use the following terminology.

By a *claw with respect to a line H*, we mean a set C of lines with the following properties

(i) $H \notin C$ and H is disjoint from each line of C.
(ii) Any two lines of C intersect.

The *order* of a claw is its number of elements.

A claw with respect to a line H in a linear space satisfying the conditions of Theorem 5.1.1, with respect to H, is called *normal* if it has order d.

A *clique* is a set of mutually disjoint lines. A clique M is *maximal* with respect to H if

(i) M is a maximal set of mutually disjoint lines, including H.
(ii) $|M| \geq n - (d - 1)c + x + 1$.

Proposition 5.1.2. Let S be as in Theorem 5.1.1. Let C be a claw of order s with respect to H. Denote by T the set of lines disjoint from H which do not lie in C. For $0 \leq y \leq s$ *let* $f(y)$ *be the number of lines in T which miss exactly y lines of C. Then* $s \leq d$. *Moreover,*

$$f(0) - \sum_{y=1}^{s} f(y)(y - 1) \geq n(d - s) + x - sc \quad \textit{and} \quad \sum_{y=0}^{s} f(y) = nd + x - s.$$

PROOF. From assumption (2) of Theorem 5.1.1 we get

$$\sum_{y=0}^{s} f(y) = |T| = nd + x - s. \tag{1}$$

Counting the set $\{(G,L) \mid G \in T, L \in C, L$ and G disjoint$\}$ in two different ways, we obtain

$$\sum_{y=0}^{s} f(y)y = \sum_{L \in C} m(L,H).$$

Together with hypothesis (3), this implies

$$\sum_{y=0}^{s} f(y)y \leq s(n-1+c) \qquad \text{and} \tag{2}$$

$$\sum_{y=0}^{s} f(y)y \geq s(n-1+a). \tag{2'}$$

Combining (1), (2) and (2′) results in one of the inequalities in the proposition. It remains to show that $s \leqq d$. For this we may assume that condition 5 (ii) is not satisfied, and hence 5 (i) holds. Let L and L' be two distinct lines of C. Since H misses L and L', there exist at most $h(L,L') - 1$ lines in T which miss L and L'. If we count the set $\{(G,L,L') \mid G \in T, L,L' \in S$, and L and L' miss $G\}$ in two different ways, we obtain

$$\sum_{y=0}^{s} f(y)y(y-1) \leq \sum_{L,L' \in C} (m(L,L') - 1),$$

and from hypothesis (4) we get

$$\sum_{y=1}^{s} f(y)y(y-1) \leq s(s-1)e.$$

Together with equations (1) and (2′) we conclude that

$$\begin{aligned} 0 &\leq f(0) + \tfrac{1}{2}\sum_{y=1}^{s} f(y)(y-1)(y-2) \\ &\leq n(d-s) + x - sa + \tfrac{1}{2}s(s-1)e. \end{aligned} \tag{3}$$

If we assume $s = d + 1$ then by equation (3) we obtain the following contradiction to hypothesis (5):

$$n = n(s-d) \leq x - sa + \tfrac{1}{2}s(s-1)e = \tfrac{1}{2}(d+1)(de-2a) + x.$$

Hence there exists no claw of order $d + 1$ and therefore also no claw with an order larger than d. This proves Proposition 5.1.2. □

Proposition 5.1.3. Let S be as in Theorem 5.1.1. For every line L which misses H there exists a normal claw with respect to H, containing L.

PROOF. Let C be a claw of maximal order s which contains L. As in Proposition 5.1.2, we denote by $f(y)$ the number of lines of C which miss H and which intersect y lines of C. Since C is a maximal claw we have $f(0) = 0$. Proposition 5.1.2 shows therefore

$$0 = f(0) \geq n(d - s) + x - sc.$$

Hence, $(n + c)s \geqq nd + x$. Assume by way of contradiction that $s < d$. In view of $n + c > 0$ (this follows from condition (3) of the theorem, since H misses some line), this implies $(n + c)(d - 1) \geqq dn + x$, i.e., $n \leqq (d - 1)c - x$. In view of condition (6), we conclude that $dc + e < x$. Hence, $n \leqq (d - 1)c - x < - c - e$. However, $n + c > 0$ and $e \geqq 0$, a contradiction. Thus, C is a normal claw containing L. □

Proposition 5.1.4. Let S be as in Theorem 5.1.1. Suppose L is a line which misses H and denote by C a normal claw with respect to H, containing L. Moreover, let M be the set of all lines $X \notin C - \{L\}$ *which miss H and which intersect every line of* $C - \{L\}$. *Then M is contained in a maximal clique.*

PROOF. Since C is a normal claw, $C' = C - \{L\}$ is a claw of order $d - 1$. Since every line of M intersects every line of C' by definition and because there is no claw of order $d + 1$, any two lines of M are disjoint. This shows that $M \cup \{H\}$ is a clique.

Proposition 5.1.2 applied to C' shows that

$$|M| = f(0) \geq n + x - (d - 1)c.$$

If M' is a maximal set of mutually disjoint lines containing $M \cup \{H\}$, then M' is therefore by definition a maximal clique. Because L is in M, this proves the proposition. □

Proposition 5.1.5. Let S be as in Theorem 5.1.1. If M_1 *and* M_2 *are distinct maximal cliques with respect to H, then* $M_1 \cap M_2 = \{H\}$.

PROOF. Because M_1 and M_2 are distinct maximal cliques, there exist intersecting lines $L_1 \in M_1$ and $L_2 \in M_2$. Every line which is in $M_1 \cap M_2$ misses L_1 and L_2. It follows therefore from hypothesis (4) that

$$|M_1 \cap M_2| = m(L_1,L_2) \leq e + 1.$$

By definition, H is a line of M_1 and M_2. Assume by way of contradiction that M_1 and M_2 have a second line L in common. Then every line in $(M_1 \cap M_2) - \{H,L\}$ misses H and L. Together with hypothesis (3) this implies

$$|M_1 \cap M_2| \leq m(L,H) + |\{L,H\}| \leq n + c + 1.$$

We obtain

$$|M_1| + |M_2| = |M_1 \cup M_2| + |M_1 \cap M_2| \leq n + c + 2 + e.$$

On the other hand, we have

$$|M_1| + |M_2| \geq 2[n - (d - 1)c + x + 1],$$

since M_1 and M_2 are maximal cliques. Combining these, we obtain a contradiction to hypothesis (6). □

PROOF OF THEOREM 5.1.1. (a) By hypothesis (2) there is a line which misses H. Proposition 5.1.3 implies therefore that there exists a normal claw C with respect to H. We shall use the notation of Proposition 5.1.2 for C. Since there is no claw of order $d + 1$, we have $f(0) = 0$. From Proposition 5.1.2 we get

$$\sum_{y=1}^{d} f(y) = nd + x - d \qquad \text{and} \qquad \sum_{y=2}^{d} f(y)(y - 1) \geq x - dc.$$

So

$$f(1) = nd + x - d - \sum_{y=2}^{d} f(y) \geq nd + x - d - \sum_{y=2}^{d} f(y)(y - 1)$$
$$\geq nd + 2x - d(c + 1).$$

Put $C = \{L_1, \ldots, L_d\}$ and let M_i' denote the set of lines $\notin C - \{L_i\}$ which miss H and which intersect every line of $C - \{L_i\}$, $1 \leq i \leq d$. Then

$$\sum_{i=1}^{d} |M_i' - \{L_i\}| = f(1).$$

By Proposition 5.1.4, the set $M_i' \cup \{H\}$ is contained in a maximal clique M_i with respect to H, for each i. By definition the set M_i' contains the line L_i but no line of $C - \{L_i\}$. Therefore the sets M_i are distinct. Assume by way of contradiction that there is another maximal clique M_{d+1}. Since any two distinct maximal cliques have only the line H in common (Proposition 5.1.5), it follows that

$$\left|\bigcup_{i=1}^{d+1}(M_i - \{H\})\right| = \sum_{i=1}^{d+1}|M_i - \{H\}| \geq f(1) + d + |M_{d+1} - \{H\}|.$$

Since every line of $M_i - \{H\}$ misses H and because the number of lines missing H is $dn + x$, we conclude that

$$dn + x \geq f(1) + d + |M_{d+1} - \{H\}| = f(1) + d + |M_{d+1}| - 1.$$

By definition, the clique M_{d+1} has at least $n - (d-1)c + x + 1$ lines. Using this and the above lower bound for $f(1)$, we obtain

$$dn + x \geq nd + 2x - d(c + 1) + d + n - (d - 1)c + x.$$

Hence $n \leqq (2d - 1)c - 2x$. This contradicts conditions (6) and (7).

Therefore there are exactly d maximal cliques. Propositions 5.1.3 and 5.1.4 show that every line which misses H is contained in exactly one of these d cliques. This proves (a).

In order to prove (b) assume now that every point outside of H has degree $n + 1$. Then every point outside of H lies on exactly d lines which miss H. Therefore every point outside of H lies on a unique line of each of the d maximal cliques. Consequently the M_i are parallel classes. □

In the following corollary we handle an important particular case.

Corollary 5.1.6. Let S be a finite linear space with a line H such that every point outside of H has the same degree $n + 1$. Denote the degree of H by $n + 1 - d$, the number of lines by $b = n^2 + n + 1 + z$ and the number of lines missing H by $dn + y + z$. Suppose there exist integers a and A with the following properties.

(1) $n + 1 - A \leqq k_L \leqq n + 1 - a$ for every line L missing H.
(2) $2n > (d + 1)(dA^2 + d - 2da + 2a - 2) - 2dy + d(d - 1)z$.
(3) $n > (2d - 1)(d - 1)(A - 1) + A^2 - 1 + (2d - 3)y + 2(d - 1)z$.

Then we have

(a) There are d parallel classes $\Pi_1, \ldots, \Pi_d$ with $H \in \Pi_i$ and such that every line missing H is contained in exactly one of the parallel classes Π_i.
(b) There is an integer t such that $|\Pi_i| \geqq t$ for all i and such that every

maximal set Π of mutually disjoint lines with $H \in \Pi$ and $|\Pi| \geqq t$ is one of the parallel classes Π_i.

PROOF. The proof will follow easily from Theorem 5.1.1.

Define $a' = (d - 1)(a - 1) + y + z$, $c = (d - 1)(A - 1) + y + z$, $e = A^2 - 1 + z$, and $x = y + z$. Then hypotheses (1) and (2) of Theorem 5.1.1 hold with a' replacing a. Since every point outside H has degree $n + 1$ and in view of property (1), it is easy to see that properties (3) and (4) of Theorem 5.1.1 also hold. The conditions (2) and (3) of the corollary just correspond to the conditions (5) and (6) in Theorem 1. Therefore the corollary follows from Theorem 5.1.1. □

In many situations, Theorem 5.1.1 or Corollary 5.1.6 will show that every line L of a linear space with maximal point degree $n + 1$ is contained in exactly $n + 1 - k_L$ maximal sets of mutually disjoint lines which have only the line L in common. In the following theorem we shall show that this information is sufficient to construct a projective plane provided that there are enough lines.

Proposition 5.1.7. Let S be a linear space with maximal point degree $n + 1$ such that the following two conditions hold.

(1) The number of lines is $b \geqq n^2$.

(2) For each line L there is an integer $t(L)$ with the following property. If $n + 1 - d$ is the degree of L, then there are exactly d sets M of mutually disjoint lines with $L \in M$ and $|M| \geqq t(L)$. Furthermore, every line missing L appears in exactly one of these d sets M.

Then S can be embedded into a projective plane of order n.

PROOF. In this proof we call a set M a *special clique* if it is a clique and if it satisfies $|M| \geqq t(L)$ for at least one line L of M. We may suppose $d \neq 0$.

For each special clique M we shall adjoin a new point to every line of M, and then we show that this extended incidence structure embeds into a projective plane. This will be done in several steps.

Step 1. A line L of degree $n + 1 - d$ is contained in exactly d special cliques.

By hypothesis (2) in Proposition 5.1.7, there are exactly d special cliques $M_1, \ldots, M_d$ with $L \in M_i$ and $|M_i| \geqq t(L)$. Furthermore, every line missing L appears in exactly one of these d cliques. Assume by way of contradiction that there exists another special clique M with $L \in M$. Then $|M| < t(L)$. By definition, M contains a line L' with $|M| \geqq t(L')$. In particular $L' \neq L$ and L' is parallel to L. Thus there is a clique M_j containing L and L'. In view of $|M_j| \geqq t(L) > |M| \geqq t(L')$, this is a contradiction to condition (2) of Proposition 5.1.7.

Step 2. If L_1 and L_2 are disjoint lines, then there exists a unique special clique containing L_1 and L_2.

This is an immediate consequence of step 1 and condition (2) of Proposition 5.1.7.

Step 3. Every point of degree $n + 1$ lies on a line of any special clique.

Let p be a point of degree $n + 1$ and let M be any special clique. We have to show that M contains a line passing through p. Let L be any line of M. If p is on L, then the proof is complete. Let us now assume that p is not on L and denote by $n + 1 - d$ the degree of L. Since p has degree $n + 1$, it lies on d lines $L_1, \ldots, L_d$ missing L. By step 2, there is a special clique M_i containing L and L_i. In this way we obtain d special cliques containing L. Step 1 shows therefore that M is one of the special cliques M_i. Therefore p lies on a line of M, which is one of the lines L_i.

Step 4. The number of special cliques is $n^2 + n + 1 - v$.

Let p be any point of degree $n + 1$; let $L_1, \ldots, L_{n+1}$ be the lines through p and denote the degree of L_i by $n + 1 - d_i$. Then

$$v - 1 = \sum_{i=1}^{n+1} (k_{L_i} - 1) = (n + 1)n - \sum_{i=1}^{n+1} d_i.$$

From step 1 it follows that

$$|\{M | M \text{ is a clique containing one of the lines } L_i\}| = \sum_{i=1}^{n+1} d_i.$$

Finally, step 3 implies that we have already counted all special cliques. The assertion follows.

We are already well prepared for the proof. For each special clique M, let p_M be a new point which we adjoin to all lines of M but to no other line of S. In this way we have extended the linear space S to an incidence structure S', which need not be a linear space. However, any two lines of S' meet in a unique point (step 2), and every line has degree $n + 1$ (step 1). Furthermore, we have now $n^2 + n + 1$ points (step 4). Since there are at least n^2 lines, it follows that S' can be extended to a projective plane P of order n (Proposition 2.2.3). But then S is also embedded in P. □

As an important consequence of Theorem 5.1.1 and Proposition 5.1.7, we obtain the following corollary.

Corollary 5.1.8. Let S be a linear space with maximal point degree $n + 1$, and suppose that the hypotheses of Theorem 5.1.1 or Corollary 5.1.6 are satisfied for every line of S. If there are at least n^2 lines, then S can be embedded into a projective plane of order n.

PROOF. We show that for every line L there exists an integer $t(L)$ satisfying hypothesis (2) of Proposition 5.1.7. If L has degree $n + 1$, we can put $t(L) = 2$. If L has degree at most n, then Theorem 5.1.1 or Corollary 5.1.6 shows that such a $t(L)$ exists. Therefore Corollary 5.1.8 follows from Proposition 5.1.7. □

Now we want to show that a linear space can be embedded into a projective plane if the maximal point degree and the minimal line degree are almost the same. First we handle the important particular case that the linear space has constant point degree. This result already has remarkable consequences.

5.2 Embedding linear spaces with constant point degree

In this section, S denotes a linear space with constant point degree $n + 1$. Then every line has degree at most $n + 1$. We call the lines of degree $n + 1$ *long* and the other lines *short*. If every line is long, then S is a projective plane of order n. We shall therefore assume throughout that there is a short line. We denote by $n + 1 - A$ the minimal degree of the short lines, and by $n + 1 - a$ the maximal degree of the short lines. The number of points is denoted by v and the number of lines by b. Furthermore, we define the integer z by $b = n^2 + n + 1 + z$. In order to use Corollary 5.1.6, we need upper and lower bounds for z. Since we have constant point degree, this is an easy task. However for linear spaces without constant point degree this will be one of the main problems.

Proposition 5.2.1. (a) We have $z \geqq -aA$.

(b) If $z > 0$ and $n > aA(A - a)$, then there is no long line and we have $A \geqq a + 2$ and $z \leqq (A - a - 1)a$.

PROOF. (a) Let L be any line of maximal degree $n + 1 - a$, and let L' be a line which intersects L in a point p. Then every point of $L' - p$ lies on a lines missing L. Thus, there are at least $(n - A)a$ lines which miss L. Since there are $(n + 1 - a)n$ lines which intersect L, it follows that there are at least $(n - A)a + 1 + (n + 1 - a)n = n^2 + n + 1 - aA$ lines.

(b) Since a long line intersects every other line and in view of $b > n^2 + n + 1$, there is no long line. Let L be a line of maximal degree $n + 1 - a$, and denote by M the set of all lines missing L. Counting incident point line pairs (p,X) with $X \in M$, we obtain

$$|M|(n + 1 - A) \leq \sum_{X \in M} k_X = \sum_{p \notin L} (r_p - k_L) = (v - k_L)a.$$

Because every line has at most $n + 1 - a$ points, we have $v \leqq k_L + n(n - a)$.

Combining these, we obtain

$$|M| \leq \frac{(v - k_L)a}{n + 1 - A} \leq \frac{n(n - a)a}{n + 1 - A}$$
$$= na + (A - a - 1)a + \frac{(A - 1)a(A - a - 1)}{n + 1 - A}.$$

Our hypothesis yields $n + 1 - A > aA(A - a) + 1 - A \geqq aA(A - a) - a(A - a) - a(A - 1) = (A - 1)a(A - a - 1)$. Therefore $|M| \leqq na + (A - a - 1)a$. It follows that

$$b = 1 + k_L n + |M| \leq n^2 + n + 1 + (A - a - 1)a,$$

i.e., $z \leqq (A - a - 1)a$. □

Theorem 5.2.2. Suppose that $b \leqq n^2 + n + 1$ and assume that S satisfies the following conditions.

(1) $n > \frac{1}{2}(A^2 - 1)(A^2 + A - 2a + 2) + \frac{1}{2}a(a - 1)z$.
(2) $n > 2(A - 1)(A^2 - A + 1) + 2(a - 1)z$.
(3) $b \geq n^2$ or $n \geq aA + 1$.

Then S can be embedded into a projective plane of order n.

PROOF. In view of condition (3) and Proposition 5.2.1(a), we have $b >$

n^2. In view of Corollary 5.1.8, it suffices to show that the hypotheses of Corollary 5.1.6 are satisfied for every line H with $k_H < n + 1$.

Consider therefore a line H of degree at most n. Define d by $k_H = n + 1 - d$, and put $y = 0$. Then $a \leqq d \leqq A$ and hypothesis (1) yields

$$\begin{aligned} n &> \tfrac{1}{2}(A^2 - 1)(A^2 + A - 2a + 2) + \tfrac{1}{2}a(a-1)z \\ &= \tfrac{1}{2}(A+1)(A^3 - 2aA + A + 2a - 2) + \tfrac{1}{2}a(a-1)z \\ &\geq \tfrac{1}{2}(d+1)(dA^2 - 2da + d + 2a - 2) + \tfrac{1}{2}d(d-1)z, \end{aligned}$$

since $z \leqq 0$. Hypothesis (2) yields

$$\begin{aligned} n &> 2(A-1)(A^2 - A + 1) + 2(a-1)z \\ &= (2A-1)(A-1)^2 + A^2 - 1 + 2(a-1)z \\ &\geq (2d-1)(d-1)(A-1) + A^2 - 1 + 2(d-1)z. \end{aligned}$$

This shows that the conditions of Corollary 5.1.6 are satisfied. □

Corollary 5.2.3. If $b \leqq n^2 + n + 1$ and $n > \frac{1}{2}(A^2 - 1)(A^2 + A - 2a + 2)$, then S can be embedded into a projective plane of order n.

If the number of lines is larger than $n^2 + n + 1$, then the linear space can clearly not be embedded into a projective plane of order n.

Theorem 5.2.4. Suppose S is a linear space with $b = n^2 + n + 1 + z > n^2 + n + 1$ lines. Then

$$\begin{aligned} 2n &\leq (A^2 - 1)(A^2 + A - 2a + 2) + A(A-1)z, \qquad \text{or} \\ n &\leq 2(A-1)(A^2 - A + 1) + 2(A-1)z. \end{aligned}$$

PROOF. Assume that our statement is false. Then, as in the proof of the last theorem, we can show that S can be embedded into a projective plane of order n. In view of $b > n^2 + n + 1$, this is not possible. □

Corollary 5.2.5. Suppose S is a linear space with constant point degree $n + 1$ and more than $n^2 + n + 1$ lines. Then $n \leqq (A^2 - 1)(A^2 + A - 2a + 2) + A(A - 1)(A - a - 1)a$.

PROOF. This follows from Theorem 5.2.4 and Proposition 5.2.1. □

5.3 Embedding linear spaces with arbitrary point degree

In this last section of the chapter we want to study linear spaces which do not have constant point degree. Our aim is to prove that a linear space with maximal point degree $n + 1$ and minimal line degree $n + 1 - a$ can be embedded in a projective plane of order n provided that n is large compared with a. Before we state the theorem and give details of the proof, we prove two propositions which will be needed.

Proposition 5.3.1. (a) For real numbers k_l, a, c with $c \leqq k_l \leqq a$ and for nonnegative integers x and y we have

$$x(c - k)^2 + y(a - k)^2 + (k_1 - k)^2 \leq \frac{b}{4}(c - a)^2,$$

where $b = x + y + 1$ and $k = (xc + ya + k_l)/b$.

(b) Let b be a positive integer. For real numbers k_i, a, c with $c \leqq k_i \leqq a$, $i = 1, \ldots, b$, we have

$$\sum_{i=1}^{b} (k_i - k)^2 \leq \tfrac{1}{4} b(c - a)^2,$$

where $k = \frac{1}{b} \Sigma_{i=1}^{b} k_i$.

PROOF. (a) We have $c \leqq k \leqq a$. If $c = k$, then the statement is trivial. Therefore we may assume that $c < k$. Without loss of generality, we may assume furthermore that $k_1 \leqq k = 0$ and $c = -1$. Then

$$-1 = c \leq k_1 = x - ya \leq k = 0 \leq a,$$

and we conclude that

$$(2k_1 + 1 - a)^2 \leq (a + 1)^2 \leq b(a - 1)^2 + (a + 1)^2,$$

implying

$$(2x - 2ya + 1 - a)^2 \leq (x + y + 1)(a - 1)^2 + (a + 1)^2.$$

So

$$4x + 4ya^2 + 4(x - ya)^2 \leq (x + y)(a + 1)^2,$$

and then

$$x + ya^2 + k_1^2 \leq \tfrac{1}{4}(x + y + 1)(a + 1)^2.$$

In view of $k = 0$ and $c = -1$, this proves part (a).

(b) We prove this statement by induction on the number u of indices

i with $k_i \notin \{a,c\}$. If $u \leqq 1$, then we may assume that every k_i, $i \geqq 2$ is equal to a or equal to c, and we can then apply part (a). Now suppose there are two distinct indices r, $s \in \{1, \ldots, b\}$ with $k_r, k_s \neq a, c$. Then we set $l_i = k_i$ for all $i \neq r, s$. Furthermore, if $k_r + k_s \leqq a + c$ (or $k_r + k_s \geqq a + c$) then we set $l_r = c (l_r = a$, resp.) and $l_s = k_r + k_s - c (l_s = k_r + k_s - a$, resp.). Then we have $(l_1 + \ldots + l_b)/b = k$ and

$$\sum_{i=1}^{b} (l_i - k)^2 = \sum_{i=1}^{b} (k_i - k)^2 + (l_r - k)^2 + (l_s - k)^2 - (k_r - k)^2 - (k_s - k)^2$$
$$\geq \sum_{i=1}^{b} (k_i - k)^2.$$

Furthermore the number of l_i with $l_i \neq a$, c is $u - 1$ or $u - 2$. The induction hypothesis used for the l_i now completes the proof. □

The following proposition can be obtained from the proof of Proposition 5.1.2.

Proposition 5.3.2. Let S be a linear space. Fix a line H and let T be a set of lines missing H. Then

$$0 \leq m_H - \sum_{L \in T} m(H,L) + \tfrac{1}{2} \sum_{K,L \in T, K \neq L} m(K,L) - \tfrac{1}{2} |T| (|T| + 1)$$

where $m_H = \{L \in \mathcal{L} | L \text{ misses } H\}$.

We devote the remainder of the chapter to proving the following result.

Theorem 5.3.3. Let S be a linear space and denote by $n + 1$ its maximal point degree and by $n + 1 - a$ its minimal line degree. If $4n > 6a^4 + 20a^3 + 51a^2$ and $a \geqq 2$, then S can be embedded in a projective plane of order n.

REMARK. The theorem remains true for $4n > 6a^4 + 9a^3 + 19a^2 + 8a$. This was proved in Beutelspacher and Metsch (1987). The proof is essentially the same as the one we give. However, in order to avoid some obscure technical details, we shall not prove the best possible bound here. See also Bruck (1963).

We want to prove Theorem 5.3.3 by applying Theorem 5.1.1 and

Proposition 5.1.7. Therefore we need bounds for the number of points and lines, and for the number of lines missing a given or two given lines. This will be done in subsequent propositions. We shall show that r, the average of the point degrees, is almost $n + 1$ and that the linear space behaves almost as a linear space with constant point degree r. Keeping this in mind helps to understand the proof. The basic tool in the proof of this theorem is Proposition 5.3.2. Recall that we used this result in the proof of Theorem 5.1.1 to show that there can not be a set of more than d mutually intersecting lines which all miss a given line H of degree $n + 1 - d$. Now we work with the average r of the point degrees, and the idea is the following. We use Proposition 5.3.2 to show that the number of lines in a set of mutually intersecting lines missing a given line H of degree $n + 1 - d$ is bounded. However the bound will not be $d = n + 1 - k_H$ as before, but something closer to $r - k_H$. Then we use the existence of a point of degree $n + 1$ to construct such a set T with d lines, and this will give us a bound for r. Having attained this bound, we shall easily be able to complete the proof.

For the proof we shall use the following notation.

$$r = \frac{1}{v}\sum_{p\in\mathcal{P}} r_p, \qquad k = \frac{1}{b}\sum_{L\in\mathcal{L}} k_L, \qquad e = n + 1 - r,$$

$$s = \frac{1}{v}\sum_{L\in\mathcal{L}} (k_L - k)^2, \qquad \text{and} \qquad t = \frac{(1 + s + k - r)(r - k)}{k}.$$

Furthermore, for every line L we define

$$d_L = n + 1 - k_L,\ t_L = \frac{1}{v - k_L}\sum_{p\notin L} r_p, \qquad e_L = r - k_L = d_L - e.$$

Finally we denote by $n + 1 - c$ the maximal line degree.

Proposition 5.3.4. We have

(a) $vr = bk$, $v = 1 + r(k - 1) + s$, $b = (r - 1)^2 + k + s + t$.
(b) $n + 1 - a \leqq k \leqq r \leqq n + 1$.
(c) $c \leqq a - 1$.

PROOF. (a) Clearly $vr = bk$. Since for every point p we have

$$v = 1 - r_p + \sum_{p\in L} k_L,$$

we conclude that

$$
\begin{aligned}
v^2 &= v - \sum_{p\in p} r_p + \sum_{p\in p}\sum_{p\in L} k_L = v - vr + \sum_{L\in\mathcal{L}} k_L^2 \\
&= v - vr + \sum_{L\in\mathcal{L}} k_L^2 - 2bk^2 - 2k\sum_{L\in\mathcal{L}} k_L \\
&= v - vr + bk^2 + \sum_{L\in\mathcal{L}} (k_L - k)^2 = v - vr + vrk + vs.
\end{aligned}
$$

Hence $v = 1 + r(k - 1) + s$ and

$$b = \frac{vr}{k} = \frac{(1 + r(k - 1) + s)r}{k} = (r - 1)^2 + k + s + t.$$

(b) Since $n + 1 - a$ is the minimal line degree and $n + 1$ is the maximal point degree, we have $n + 1 - a \leqq k$ and $r \leqq n + 1$. The equation $k \leqq r$ follows from $vr = bk$ and $b \geqq v$.

(c) Assume by way of contradiction that $c > a - 1$. Since $n + 1 - a$ is the minimal and $n + 1 - c$ is the maximal line degree, it follows that every line has the same degree, which is $n + 1 - a$. Hence, every point has the same degree, which is $n + 1$, since $n + 1$ is the maximal point degree. It follows that $v = 1 + (n + 1)(n - a)$ and $b(n + 1 - a) = v(n + 1)$. Hence, $0 \equiv v(n + 1) \equiv (1 - a)a$ (modulo $n + 1 - a$). In view of $a \geqq 2$ and $n \geqq a^4$, this is not possible. □

Proposition 5.3.5. We have $s + t < \frac{1}{4}a^2 + \frac{5}{32}$ and $|t| < \frac{1}{8}$.

PROOF. From Proposition 5.3.1(b) we get

$$s = \frac{1}{v}\sum_{L\in\mathcal{L}} (k_L - k)^2 \leq \frac{b}{4v}a^2 = \frac{r}{4k}a^2 \leq \frac{n + 1}{4(n + 1 - a)}a^2 < \tfrac{1}{4}a^2 + \frac{1}{32}.$$

This implies

$$
\begin{aligned}
|t| &= \frac{|1 + s + k - r|(r - k)}{k} \leq \frac{(1 + s + a)a}{k} \\
&< \frac{\left(\frac{7}{6} + \frac{a^2}{4} + a\right)a}{n + 1 - a} \leq \tfrac{1}{8}.
\end{aligned}
$$

□

Proposition 5.3.6. For every line L we have

$$t_L - k_L \le e_L + \frac{a}{k-2},\ t_L - k_L \ge e_L - \frac{a}{n-a},\ k_L \le t_L, \qquad \text{and}$$

$$-1 < e_L \le d_L.$$

PROOF. Since every point outside the line L has degree at least k_L we have $t_L \geqq k_L$. Now define

$$y = \frac{1}{k_L} \sum_{p \in L} r_p.$$

Then y is the average of the point degrees of L. Since t_L is the average of the degrees of points outside of L, we obtain

$$vr = (v - k_L)t_L + k_L y$$

and

$$r - t_L = \frac{k_L(v - t_L)}{v}.$$

Since there is a point of degree $n + 1$, we have $v \geqq 1 + (n + 1)(n - a) \geqq k_L\,(n - a)$, and because y and t_L, as the average of point degrees, lie between $n + 1 - a$ and $n + 1$, we have $|y - t_L| \leqq a$. Hence,

$$|r - t_L| \le \frac{k_L a}{v} \le \frac{a}{n-a}.$$

In particular $k_L \leqq t_L \leqq r + a/(n - a)$ and it follows that

$$\begin{aligned}|r - t_L| &\le \frac{k_L a}{v} \le \frac{k_L a}{r(k-1)} \le \frac{a}{k-1} + \frac{a^2}{r(k-1)(n-a)}\\ &\le \frac{a}{k-1} + \frac{a^2}{(k-2)(k-1)(n-a)}\\ &\le \frac{a}{k-1} + \frac{a}{(k-1)(k-2)} = \frac{a}{k-2}.\end{aligned}$$

The stated bounds for $t_L - k_L$ follow now from $t_L - k_L = t_L - r + e_L$.

Finally, $-1 \leqq e_L \leqq d_L$ follows from $e_L = r - k_L \leqq n + 1 - k_L = d_L$ and

$$e_L = r - k_L \geq r - t_L \geq -\frac{a}{k-2} \geq -1.$$

□

Proposition 5.3.7. For every line L we have

$$(n-a)e_L - a - \tfrac{1}{6} < m_L < (r-1)e_L + (a-c)a + \tfrac{1}{3}.$$

PROOF. Let M be the set of lines parallel to L. Then $m_L = |M|$ and

$$\sum_{X \in M} k_X = \sum_{p \notin L} (r_p - k_L) = (v - k_L)(t_L - k_L).$$

The lines parallel to L have degree at least $n + 1 - a$, since $n + 1 - a$ is the minimal line degree, and they have degree at most n, since a line of degree $n + 1$ meets every other line. It follows that

$$(n+1-a)m_L \leq (v - k_L)(t_L - k_L) \leq m_L(n+1-c) \leq m_L n.$$

Now we use upper and lower bounds for the number v of points to obtain bounds for m_L. Since there is a point of degree $n + 1$, we have

$$\begin{aligned} v - k_L &\geq 1 + (n+1)(n-a) - k_L \\ &\geq 1 + (n+1)(n-a-1) = n(n-a) - a. \end{aligned}$$

Hence

$$\begin{aligned} m_L &\geq \frac{[n(n-a)-a](t_L - k_L)}{n} = (n-a)(t_L - k_L) - \frac{a(t_L - k_L)}{n} \\ &\geq (n-a)(t_L - k_L) - \tfrac{1}{6} \\ &\geq (n-a)e_L - a - \tfrac{1}{6} \end{aligned}$$

(see Proposition 5.3.6).

Now we use

$$\begin{aligned} v - k_L &= 1 + r(k-1) + s - k_L \leq 1 + r(k-2) + s + a \\ &\leq r(k-2) + a^2, \qquad \text{if } a > 2, \end{aligned}$$

to obtain an upper bound for m_L:

$$m_L \le \frac{(v-k_L)(t_L-k_L)}{n+1-a} \le \frac{[r(k-2)+a^2](t_L-k_L)}{n+1-a}$$

$$\le \frac{r(k-2)(t_L-k_L)+a^3}{n+1-a}$$

$$\le \frac{r(k-2)\left(e_L+\dfrac{a}{k-2}\right)+a^3}{n+1-a} \qquad \text{(Proposition 5.3.6)}$$

$$\le \frac{r(k-2)e_L+ra+a^3}{n+1-a} \le \frac{r(n-1-c)e_L+ra+a^3}{n+1-a}$$

$$\le re_L + \frac{r(a-2-c)e_L+ra+a^3}{n+1-a}$$

$$\le re_L + \frac{(n+1)(a-2-c)e_L+(n+1)a+a^3}{n+1-a}$$

$$= re_L + (a-2-c)e_L + a + \frac{a(a-2-c)e_L+a^2+a^3}{n+1-a}$$

$$\le (r-1)e_L + (a-1-c)e_L + a + \tfrac{1}{3}.$$

We leave the reader to verify that this inequality also holds for $a = 2$. In view of $c \leqq a - 1$, we have $(a-1-c)e_L \leqq (a-1-c)a$, and this completes the proof. □

Proposition 5.3.8. For any lines H and L which have degree at most n, we have

$$n - \tfrac{5}{4}a^2 - 4a \le m_H + m_L - b + k_Hk_L \le n + 3a^2 - a - 2ac.$$

PROOF. Define $y = m_H + m_L - b + k_Hk_L$. In view of $b = (r-1)^2 + k + s + t$, we have

$$y = m_H + m_L - b + (r-e_H)(r-e_L)$$
$$= m_H + m_L + 2r - 1 - k - s - t - r(e_H + e_L) + e_He_L.$$

Since $2r - 1 - e_H - e_L = 2(n+1-e) - 1 - (d_H - e) - (d_L - e) = 2n + 1 - d_H - d_L$, this implies

$$y = m_H - (r-1)e_H + m_L - (r-1)e_L + 2n + 1 - k - s - t - (d_H + d_L) + e_He_L.$$

Together with the upper bound for m_H and m_L of Proposition 5.3.7, we obtain

$$y < 2[(a-c)a + \tfrac{1}{3}] + 2n + 1 - k - s - t - (d_H + d_L) + e_H e_L.$$

In view of $n + 1 - k \leqq a$, $|t| < \frac{1}{8}$ by Proposition 5.3.5, and $s \geqq 0$, this yields

$$y < 2(a-c)a + a + 1 + n - (d_H + d_L) + e_H e_L.$$

By the assumption of Proposition 5.3.8, we have $d_H \geqq 1$ and $d_L \geqq 1$, and by Proposition 5.3.7 we have $-1 \leqq e_H \leqq d_H$ and $-1 \leqq e_L \leqq d_L$. Therefore we have $e_H e_L \leqq d_H d_L$. It follows that

$$\begin{aligned} y &< 2(a-c)a + a + 1 + n - (d_H + d_L) + d_H d_L \\ &\leq (a-c)2a + a + 1 + n - 2a + a^2 \\ &= n + 3a^2 - a + 1 - 2ac. \end{aligned}$$

Now we use the lower bound for m_H and m_L of Proposition 5.3.7 to obtain

$$\begin{aligned} y > (n + 1 - r - a)(e_H + e_L) - 2a - \tfrac{1}{2} + 2n \\ + 1 - k - s - t - (d_H + d_L) + e_H e_L. \end{aligned}$$

Since $n + 1 - r = e$, $n + 1 - k \geqq 0$, and $s + t < \frac{1}{4}a^2 + \frac{5}{32}$ by Proposition 5.3.5, it follows that

$$y > (e-a)(e_H + e_L) - 2a - \tfrac{1}{2} + n - \tfrac{1}{4}a^2 - \tfrac{1}{2} - (d_H + d_L) + e_H e_L.$$

Now we use

$$\begin{aligned} (e-a)(e_H + e_L) + e_H e_L &= (e_H + e)(e_L + e) - e^2 - a(e_H + e_L) \\ &= d_H d_L - e^2 - a(d_H + d_L - 2e) \\ &\geq d_H d_L - a(d_H + d_L) \geq -a^2 \end{aligned}$$

and obtain

$$y > n - \tfrac{5}{4}a^2 - 2a - 1 - (d_H + d_L) \geq n - \tfrac{5}{4}a^2 - 4a - 1.$$

□

Proposition 5.3.9. (a) If G and L are intersecting lines, then $m(G,L) \leqq 3a^2 + a - 2ac$.

(b) If G and L are disjoint lines, then

$$n - \tfrac{5}{4}a^2 - 4a \leq m(G,L) \leq n + 3a^2 - a.$$

PROOF. (a) If one of the lines G and L has degree $n + 1$, then there is no line missing it and the statement is trivial. Therefore we assume that G and L have degree at most n. Let x be the number of lines intersecting G and L, and denote by p the point of intersection of G and L. Then $b = x + m_G + m_L - m(G,L)$ and

$$\begin{aligned} x &= (k_G - 1)(k_L - 1) + r_p = k_G k_L - k_G - k_L + 1 + r_p \\ &\leq k_G k_L - 2(n + 1 - a) + 1 + n + 1 = k_G k_L - n + 2a. \end{aligned}$$

Thus

$$m(G,L) = m_G + m_L - b + x \leq m_G + m_L - b + k_G k_L - n + 2a,$$

and the statement follows from Proposition 5.3.8.

(b) If x is the number of lines meeting G and L, then $b = x + m_G + m_L - m(G,L)$ and $x = k_G k_L$. Therefore $m(G,L) = m_G + m_L - b + k_G k_L$, and the statement follows from Proposition 5.3.8 (notice that G and L have degree at most n, since they are disjoint). □

Proposition 5.3.10. $b > n^2 + 1$ *and* $4en < 2a^4 + 5a^3 + a^2 + 4$.

PROOF. Let H be any line such that there is a point p of degree $n + 1$ outside of H. Furthermore, set $d = d_H$, and denote by T the set of the d lines through p which miss H. Then from Proposition 5.3.2 and Proposition 5.3.9 we obtain

$$\begin{aligned} 0 &\leq m_H - \sum_{L \in T} m(H,L) + \tfrac{1}{2} \sum_{G,L \in T, G \neq L} m(G,L) \\ &\leq m_H - d(n - \tfrac{5}{4}a^2 - 4a - 1) + \tfrac{1}{2}d(d-1)(3a^2 + a - 2ac) \\ &\leq (r - 1)e_H + a^2 + 1 - d(n - \tfrac{5}{4}a^2 - 4a - 1) \\ &\quad + \tfrac{1}{2}d(d-1)(3a^2 + a - 2ac). \end{aligned}$$

In view of $(r - 1)e_H - dn = (n - e)(d - e) - dn = e^2 - ed - en$, it follows that

$$\begin{aligned} en &\leq e(e - d) + a^2 + 1 + d(\tfrac{5}{4}a^2 + 4a + 1) \\ &\quad + \tfrac{1}{2}d(d-1)(3a^2 + a - 2ac). \end{aligned} \tag{4}$$

Now $d \leqq a$ and $c \geqq 0$ yield

$$\begin{aligned} en &\leq a^2 + 1 + a(\tfrac{5}{4}a^2 + 4a + 1) + \tfrac{1}{2}a(a-1)(3a^2 + a) \\ &= \tfrac{1}{4}(6a^4 + a^3 + 18a^2 + 4a + 4) < n - a. \end{aligned}$$

We wish to apply equation (4) now to the special case $d = c$. In order to do this, we have to show that outside of some line of maximal degree $n + 1 - c$ there is a point of degree $n + 1$. We shall show this by proving that there are at least $n + 2$ points of degree $n + 1$.

Assume to the contrary that there are at most $n + 1$ points of degree $n + 1$. Then

$$v(n + 1 - e) = vr = \sum_{p \in p} r_p \leq vn + n + 1,$$

so that $v(1 - e) \leqq n + 1$. In view of $v > (n + 1)(n - a)$, we obtain $(n - a)(1 - e) < 1$. Since $en < n - a$, we obtain $n > (n - a)(n - en) \geqq (n - a)a$, which is a contradiction. Therefore we can replace d with c in equation (4). Since it follows from the above that $e < 1$, we can replace the term $e(e - d)$ in equation (4) by 1 and obtain

$$\begin{aligned} 4en &\leq 4 + 4a^2 + 4 + c(5a^2 + 16a + 4) + 2c(c - 1)(3a^2 + a - 2ac) \\ &= -4ac^3 + (6a^2 + 6a)c^2 + (-a^2 + 14a + 4)c + 4a^2 + 8. \end{aligned}$$

Since as functions of c, $-4ac^3 + 6a^2c^2$ and $6ac^2 - a^2c$ obtain their maximum value at $a - 1$ in the interval $[0, a - 1]$, and in view of $0 \leqq c \leqq a - 1$, we can replace c by $a - 1$ in the above inequality and obtain

$$\begin{aligned} 4en &\leq -4a(a - 1)^3 + (6a^2 + 6a)(a - 1)^2 \\ &\quad + (-a^2 + 14a + 4)(a - 1) + 4a^2 + 8 \\ &= 2a^4 + 5a^3 + a^2 + 4 < \frac{4n}{3}. \end{aligned}$$

Therefore $e < \frac{1}{3}$ and $r = n + 1 - e > n + \frac{2}{3}$. Together with Proposition 5.3.4 it follows that

$$\begin{aligned} b &= (r - 1)^2 + k + s + t \geq (n - \tfrac{1}{3})^2 + k + t \\ &\geq (n - \tfrac{1}{3})^2 + n - a > n^2 + 1. \end{aligned}$$

□

Proposition 5.3.11. Every line H which has degree at most n satisfies the hypothesis of Theorem 5.1.1.

PROOF. Let $n + 1 - d$ be the degree of H and set $x = m_H - dn$. We have to show that there are integers α, y, and ϵ such that the following conditions are fulfilled (see Theorem 5.1.1).

(1) $dn + x > 0$.
(2) $n + 1 - \alpha \leqq m(H,L) \leqq n - 1 + y$ for every line L missing H.
(3) $m(G,L) \leqq \epsilon + 1$ for any two intersecting lines G and L missing H.
(4) $2n > (d + 1)(d\epsilon - 2\alpha) + 2x$.
(5) $n > (2d - 1)y + \epsilon - 2x$.
(6) $\epsilon \geqq 0$.

We shall prove that these conditions are met if we define

$$\alpha = -\frac{5}{4}a^2 - 4a,\ y = 3a^2 - a + 1,\text{ and } \epsilon = 3a^2 + a - 1.$$

First notice that conditions (2) and (3) follow immediately from Proposition 5.3.9 and that condition (6) is fulfilled. In order to show that the other three conditions are satisfied, we need a lower and an upper bound for x. Using Proposition 5.3.7, we get

$$\begin{aligned} x = m_H - dn &\geq (n - a)e_H - a - \tfrac{1}{4} - dn \\ &= n(e_H - d) - ae_H - a - \tfrac{1}{4} \geq n(e_H - d) - a^2 - a - \tfrac{1}{4} \\ &= -en - a^2 - a - \tfrac{1}{4} \end{aligned}$$

and

$$x \leq (r - 1)e_H + a^2 + \tfrac{1}{3} - dn < a^2 + 1.$$

Using the bound for en in Proposition 5.3.10 and the hypothesis for n of Theorem 5.3.3, we obtain $dn + x > 0$ so that condition (1) is satisfied. Now we show that condition (4) holds:

$$\begin{aligned} (d + 1)(d\epsilon - 2\alpha) + 2x &\leq (a + 1)(a\epsilon - 2\alpha) + 2x \\ &\leq (a + 1)[a(3a^2 + a - 1) + \tfrac{5}{2}a^2 + 8a] + 2a^2 + 2 \\ &= (a + 1)(3a^3 + \tfrac{7}{2}a^2 + 7a) + 2a^2 + 2 \\ &\leq 3a^4 + \tfrac{13}{2}a^2 + \tfrac{25}{2}a^2 + 7a + 2 < 2n. \end{aligned}$$

This is condition (4). Finally we prove condition (5) using the bound for $2en$ of Proposition 5.3.10.

$$\begin{aligned} (2d - 1)y + \epsilon - 2x &\leq (2a - 1)(3a^2 - a + 1) + (3a^2 + a - 1) \\ &\quad + 2(en + a^2 + a + \tfrac{1}{4}) \\ &= 6a^2 + 6a - \tfrac{3}{2} + 2en \\ &\leq 6a^3 + 6a - \tfrac{3}{2} + a^4 + \tfrac{5}{2}a^3 + \tfrac{1}{2}a^2 + 2 \\ &= a^4 + \tfrac{17}{2}a^3 + \tfrac{1}{2}a^2 + 6a + \tfrac{1}{2} < n. \end{aligned}$$

This completes the proof of the proposition. □

It follows now from Propositions 5.3.10, 5.3.11 and 5.1.7 that S can be embedded into a projective plane of order n.

5.4 Exercises

1. Prove Proposition 5.3.2.
2. Complete the missing case $a - 2$ of Proposition 5.3.7.

5.5 Research problems

1. Show that the bound on n in Corollary 5.2.3 is the best possible or, alternatively, improve it.

5.6 References

Beutelspacher, A. and Metsch, K. (1986), Embedding finite linear spaces in projective planes. *Ann. Discrete Math. 30* 39–56.

Beutelspacher, A. and Metsch, K. (1987), Embedding finite linear spaces in projective planes II. *Ann. Discrete Math. 66* 219–230.

Bruck, R. H. (1963), Finite Nets II: Uniqueness and Imbedding. *Pacific J. Math. 13* 421–457.

6

Linear spaces with few lines

6.1 Restricted linear spaces

By the Fundamental Theorem (see Theorem 1.5.5) we know that $b \geqq v$ for any finite linear space. Moreover, the linear spaces with $b = v$ are precisely the (possibly degenerate) projective planes. It is natural to ask "What are the finite linear spaces satisfying $b = v + 1$, $b = v + 2$, . . .?" Indeed, these questions have been asked and answered in J. Totten (1975, 1976a, b).

In this chapter we want, however, to consider a much more general situation. Following Totten, we call a finite linear space *restricted* if it satisfies $(b - v)^2 \leqq v$. The aim of this chapter is to classify all restricted linear spaces.

In order to formulate the results one particular notion is important. Let A be an affine plane, and denote by $\Pi_1, \ldots, \Pi_m$ some of its parallel classes. Denote by S' any linear space with point set $\{\Pi_1, \ldots, \Pi_m\}$. Let S be the linear space

whose *points* are the points of A and the sets $\Pi_1, \ldots, \Pi_m$, and whose *lines* are the lines of A and the lines of L.

Then S is called an affine plane with the linear space S' *at infinity*.

For example, the *Fano quasi-plane* is the affine plane of order 2 with a near pencil at infinity (see Figure 6.1.1). (The affine plane of order 2 consists of the three "vertices" and the "midpoint" of the triangle.) Now we can formulate the main result of this chapter.

Theorem 6.1.1. Let S be a finite linear space with v points and b lines. If S is restricted, then S is one of the following structures:

(a) a generalized projective plane,

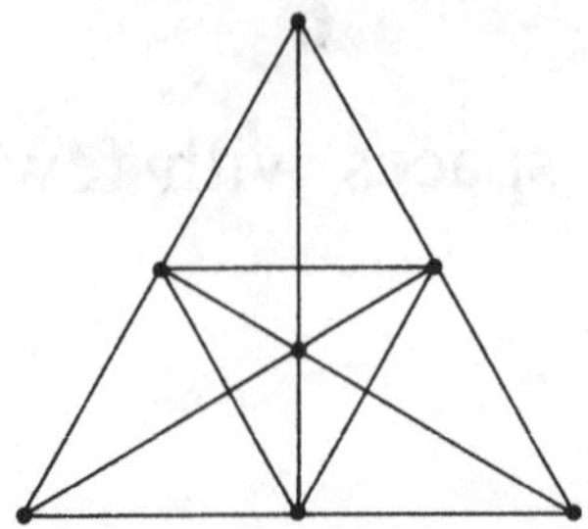

Figure 6.1.1. The Fano quasi-plane.

(b) *an affine plane, a (possibly punctured) affine plane with one point at infinity, or a projective plane of order n with all its lines, but at most n points removed,*
(c) *an affine plane with a generalized projective plane P at infinity,*
(d) *the (3,4)-cross.*

In fact, a little bit more is true. One can relax the hypotheses of Theorem 6.1.1 slightly and get the following characterization (see Metsch 1991).

Result 6.1.2. Let S be a finite linear space with v points and b lines satisfying $(b - v)^2 \leqq b$. *Then S is one of the following structures:*
(a) *a generalized projective plane,*
(b) *an affine plane, a (possibly punctured) affine plane with one point at infinity or a projective plane of order n with all its lines, but at most n points removed,*
(c) *an affine plane with a generalized projective plane at infinity,*
(d) *an affine plane of order n with a punctured projective plane of order m at infinity satisfying* $m^2 + m \in \{n, n + 1\}$,
(e) *a punctured affine plane of order n with a generalized projective plane P at infinity, where P has n + 1 points,*
(f) *the (3,4)-cross, the (3,5)-cross or the linear space shown in Figure 6.1.2.*

We cannot overemphasize the importance of Theorem 6.1.1. In many situations, linear spaces with a small number of points occur; often they satisfy the hypothesis of the main theorem and can therefore be described precisely.

The proof of Theorem 6.1.1 is the aim of this chapter. It will follow in a series of propositions.

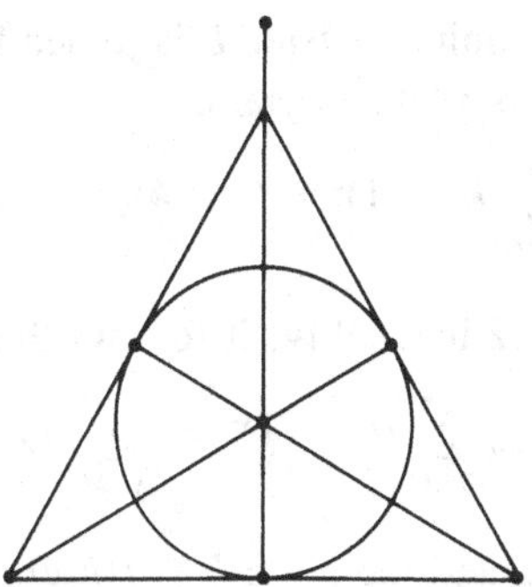

Figure 6.1.2. An exceptional linear space with $v = 8$. (Lines of size 2 are not drawn.)

From now on, S denotes a finite linear space satisfying $(b - v)^2 \leqq v$. Let n be the uniquely defined positive integer with

$$n^2 - n + 1 = (n - 1)^2 + (n - 1) + 1 < v \leqq n^2 + n + 1.$$

Since $(b - v)^2 \leqq v < (n + 1)^2$, we have $b \leqq v + n$.
We define the function f by

$$f(k,v) = 1 + k^2(v - k)/(v - 1).$$

Proposition 6.1.3. (Stanton and Kalbfleisch 1972.)

(a) If L is a line of S, then $b \geqq f(k_L,v)$.
(b) If $2 \leqq c \leqq k \leqq d \leqq v - 1$, then

$$f(k,v) \geqq \min\{f(c,v), f(d,v)\}.$$

PROOF. (a) Consider the set M of lines that intersect L in just one point and define k to be k_L. We show that M contains at least $f(k,v) - 1$ lines. Since every point of L is joined to the points outside L by a line of M, we have

$$\sum_{X \in M} (k_X - 1) = k(v - k).$$

Counting in two ways the triples (p,q,X), where X is a line of M and p, q are points on X with $p, q \neq X \cap L$, we obtain

$$\sum_{X \in M} (k_X - 1)(k_X - 2) \leq (v - k)(v - k - 1),$$

since each pair (p,q) of points outside L is joined by at most one line of M. Combining these two relations yields

$$\sum_{X\in M} (k_X - 1)^2 \leq (v - k)(v - 1).$$

By the Cauchy–Schwartz inequality, it follows that

$$|M|(v - k)(v - 1) \geq |M| \sum_{X\in M} (k_X - 1)^2 \geq (\sum_{X\in M} (k_X - 1))^2 = k^2(v - k)^2.$$

Hence $|M| \geqq f(k,v) - 1$ and therefore $b \geqq f(k,v)$.

(b) We fix v and consider k as a real variable. Then f has the derivative $(2kv - 3k^2)/(v - 1)$. Hence f has only one maximum in the interval $[2, v - 1]$. □

Proposition 6.1.4. If S contains a line with more than $n + 1$ points, then S is either a near-pencil, or the (3,4)-cross.

PROOF. Let L be a line of maximum degree k of S. By our hypothesis, $k \geqq n + 2$. Let us suppose that S is not a near-pencil. Then $n + 2 \leqq k \leqq v - 2$, in particular $v \geqq n + 4$. We claim that $n = 2$.

Indeed, by Proposition 6.1.3, we have $b \geqq \min\{f(n + 2,v), f(v - 2, v)\}$. We shall distinguish two cases.

If $b \geqq f(v - 2,v)$ then $b > 2v - 5$. Since $b \leqq v + n$, it follows that $v \leqq n + 4$. Since $v \geqq n^2 - n + 2$, this implies $n = 2$.

On the other hand, if $b \geqq f(n + 2,v)$, then we have

$$v + n \geqq b \geqq f(n + 2,v) = 1 + (n + 2)^2(v - n - 2)/(v - 1).$$

Therefore

$$\begin{aligned} 0 &\leqq v^2 - v(n^2 + 3n + 6) + n^3 + 6n^2 + 11n + 9 \\ &= (n^2 + n + 1 - v)(n + 5 - v) + 5n + 4 - vn. \end{aligned}$$

If $v \geqq n + 5$, then $vn \leqq 5n + 4$, hence $(n + 5)n \leqq 5n + 4$ and therefore $n = 2$.

So we have $n = 2$, $6 = n + 4 \leqq v \leqq n^2 + n + 1 = 7$, and $b \leqq v + 2$. Since there is a line of degree at least $n + 2 = 4$, it follows easily that S is the (3,4)-cross. □

From now on we shall assume that every line has degree at most $n + 1$. In the following proposition we show that 'in general' all points have large degree.

Proposition 6.1.5. Suppose that S has a point p_0 of degree at most n. Then one of the following possibilities occurs.

(a) S is a (possibly punctured) affine plane with a point at infinity.
(b) S is a near-pencil with four points.

PROOF. Since each line of S has at most $n + 1$ points, we have

$$v \leqq 1 + r_{p_0} \cdot n \leqq n^2 + 1.$$

Since $v \geqq n^2 - n + 2$ we have $r_{p_0} = n$. Since S is restricted it follows that $b \leqq v + n \leqq n^2 + n + 1$. We distinguish two cases.

Case 1. The point p_0 is on at least two $(n + 1)$-lines, say L_1 and L_2.

Then any point different from p_0 has degree at least $n + 1$. Hence the line L_1 intersects at least $n - 1 + n \cdot n = n^2 + n - 1$ other lines. Therefore $b \geqq n^2 + n$.

We claim that every point different from p_0 has degree $n + 1$. To see this, assume that there exists a point p whose degree is bigger than $n + 1$. Without loss of generality we may suppose that $p \notin L_1$. Then p is on a line H parallel to L_1. Since there is a line disjoint to L_1, we have $b \geqq n^2 + n + 1$. Since

$$b \leqq v + n \leqq n^2 + 1 + n$$

it follows that $b = n^2 + n + 1$ and $v = n^2 + 1$. Therefore, any line through the n-point p_0 has degree $n + 1$. In particular, pp_0 would be an $(n + 1)$-line and would hence be intersected by at least $n - 1 + (n - 1)n + r_p \geqq n^2 + n$ other lines. So, there would not exist a line disjoint to pp_0, a contradiction since every point on H has degree at least $n + 2$.

We conclude that every point different from p_0 has degree $n + 1$. So there is no line disjoint to L_1, which implies $b = n^2 + n$. Hence $v = n^2$ or $v = n^2 + 1$.

If $v = n^2 + 1$, then every line through p_0 has degree $n + 1$, while any other line has degree n. In other words, S is an affine plane of order n with a point at infinity. If $v = n^2$, then there is precisely one line of degree n through p_0. This line N is contained in a unique parallel class Π. If we adjoin Π to S, we get again an affine plane with one point at infinity.

Case 2. The point p_0 is on at most one $(n + 1)$-line.

Since p_0 has degree n and since $v \geqq n^2 - n + 2$, it follows that there is (precisely) one $(n + 1)$-line L through p_0, while any other line through p_0 must have degree n. Consequently, $v = n^2 - n + 2$. We want to show that $n = 2$ (so that S is the near-pencil on 4 points). Assume to the contrary that $n \geqq 3$. Then there are two n-lines N_1 and N_2 through p_0. Since any point different from p_0 of $N_1 \cup N_2$ has at least degree $n + 1$, it follows that $b \geqq n^2 + n - 1$; hence $b - v \geqq n^2 + n - 1 - (n^2 - n + 2) = 2n - 3$. Thus

$$(2n - 3)^2 \leqq (b - v)^2 \leqq v = n^2 - n + 2$$

or $3n^2 - 11n + 7 \leqq 0$, contradicting $n \geqq 3$. □

From now on we shall suppose that any point of S has at least degree $n + 1$.

Proposition 6.1.6. If any line has at most n points, then S is an affine plane of order n.

PROOF. We define the *deficiency* d_p of a point p as $d_p = r_p - (n + 1)$. Similarly, the *deficiency* of the line L is $d_L = n + 1 - k_L$. It follows that $d_p \geqq 0$ for any point p, and $d_L \geqq 1$ for any line L. Moreover, we have

$$\sum_{p \in \mathcal{P}} d_p + \sum_{L \in \mathcal{L}} d_L = \sum_{p \in \mathcal{P}} r_p - v(n + 1) - \sum_{L \in \mathcal{L}} k_L + b(n + 1)$$
$$= (b - v)(n + 1).$$

In particular,

$$\sum_{p \in \mathcal{P}} d_p \leqq (b - v)(n + 1) - b = (b - v)n - v \leqq n \cdot n - v < v.$$

Hence there is a point q with $d_q = 0$ and $r_q = n + 1$. Let us denote by c the number of n-lines through q. Then

$$n^2 - n + 1 \leqq v - 1 \leqq c(n - 1) + (r_q - c)(n - 2)$$
$$= n^2 - n - 2 + c \leqq n^2 - 1.$$

Consequently, $v \leqq n^2$ and $c \geqq 3$. Consider two distinct n-lines N_1 and N_2 through q. Since there are at least $n^2 + n$ lines intersecting N_1 or N_2, $b \geqq n^2 + n$. Since $(b - v)^2 \leqq v \leqq n^2$, necessarily $b = n^2 + n$ and $v = n^2$. Moreover it follows that

$$b \leqq \sum_{p \in \mathcal{P}} d_p + b \leqq \sum_{p \in \mathcal{P}} d_p + \sum_{L \in \mathcal{L}} d_L = (b - v)(n + 1) = n(n + 1) = b.$$

Consequently, $d_p = 0$ for every point p and $d_L = 1$ for every line L. Therefore, S is an affine plane. □

From now on, we shall suppose that there exists a line of degree $n + 1$ in S.

Proposition 6.1.7. (a) Any two $(n + 1)$-lines intersect in a common point.
(b) Every $(n + 1)$-line has a point of degree $n + 1$.

PROOF. Since every point has degree at least $n + 1$ and since $b \leqq n^2 + 2n + 1$, it is not possible that every point on an $(n + 1)$-line has degree bigger than $n + 1$. This implies (a) and (b). □

Proposition 6.1.8. If $b \leqq n^2 + n + 1$, then S can be obtained from a projective plane of order n by removing at most n points.

PROOF. Since every point has at least degree $n + 1$ and since there is an $(n + 1)$-line L, it follows that $b \geqq n^2 + n + 1$. Hence $b = n^2 + n + 1$. This implies in turn that every point on L has degree $n + 1$ and that there is no line disjoint to L. Thus also every point off L is an $(n + 1)$-point.

Since S is restricted, $v \geqq b - n = n^2 + 1$. So, by Proposition 2.2.3, S is embeddable in a projective plane of order n. □

From now on, we may suppose that $b \geqq n^2 + n + 2$. Let us list all the assumptions we have made so far.

- Every line has degree at most $n + 1$ and there exists a line of degree $n + 1$.
- Every point has degree at least $n + 1$ and there exists a point of degree at least $n + 2$. Note that if all points have degree $n + 1$, then every line meets an $(n + 1)$-line, and so $b = n^2 + n + 1$.
- $v \geqq n^2 + 2$ and $b \leqq v + n \leqq n^2 + 2n + 1$. (This follows from $v + n \geqq b \geqq n^2 + n + 2$.)

In order to complete the proof of Totten's theorem, we have to show that S is an affine plane of order n with a generalized projective plane at infinity. We shall do this twice, demonstrating two essentially different methods.

6.2. The combinatorial approach

This approach follows that of J. Totten, though we shall considerably simplify some of his arguments.

We shall call a point *real* if it has degree $n + 1$ and the points of degree at least $n + 2$ will be called *ideal*. A line is called *real* if it meets every line of degree $n + 1$, *ideal* if it misses some line of degree $n + 1$, and *hyperideal* if it misses all lines of degree $n + 1$. Also we call a line *long* if it has degree $n + 1$.

The following assertion is crucial for our purposes. It plays the role of Theorem 4 in Totten (1976a).

Proposition 6.2.1. There exists a long line having at least one ideal point.

PROOF. Assume to the contrary that there is a linear space fulfilling all our hypotheses but in which every long line has only real points. Among all such counterexamples (with at most $n^2 + n + 1$ points), let S be a linear space with the maximum number of points.

The following easy observation is crucial in this proof. If p is a point outside two lines L and L' and if t denotes the number of lines through p that miss L and meet L' then $t + k_L - k_{L'}$ is the number of lines through p that miss L' and meet L.

Consider a long line L. Since (outside L) there exists an ideal point which does not lie on any long line, not every line meeting L is long. Among all lines that meet L and do not have degree $n + 1$, we choose a line N of maximum degree and define $k_N = n + 1 - d < n$. We consider the set $\mathcal{X}$ consisting of all lines that miss L and meet N, and the set $\mathcal{M}$ of all lines that miss N and meet L. Since every point on L has degree $n + 1$, we have $|\mathcal{M}| = dn$. For a point p outside $N \cup L$ we denote by x_p the number of lines of $\mathcal{X}$ through p. Then $d + x_p$ is the number of lines of $\mathcal{M}$ through p. It follows that

$$\sum_{X \in \mathcal{X}} (k_X - 1) = \sum_{p \notin N \cup L} x_p$$

and

$$\sum_{M \in \mathcal{M}} (k_M - 1) = \sum_{p \notin N \cup L} (x_p + d).$$

Since $|\mathcal{M}| = dn$, we conclude that

$$(v - k_N)d + \sum_{X \in \mathcal{X}} (k_X - 1) = (v - k_N)d + \sum_{p \notin N \cup L} x_p = \sum_{M \in \mathcal{M}} k_M \qquad (1)$$

Since N and L meet in a point of degree $n + 1$, no line of $\mathcal{M}$ has degree $n + 1$. Hence $k_M \leqq k_N$ for all lines $M \in \mathcal{M}$. Using equation (1) and

$v - k_N > n^2 - n$, we obtain $d = 1$. Thus N has degree n and $|\mathcal{M}| = n$. Equation (1) implies

$$v - n + |\mathcal{X}| \leq \sum_{M \in \mathcal{M}} k_M. \tag{2}$$

Furthermore,

(a) If N_1 and N_2 are parallel n-lines that meet L, then there is no line which misses L and meets N_1 and N_2.

This can be seen as follows: We may suppose that $N = N_1$. Assume that N_2 meets a line $X \in \mathcal{X}$ in a point q. Then $x_q \geqq 1$. Since q lies on $1 + x_q$ lines of $\mathcal{M}$, there is a line $G \neq N_2$ of $\mathcal{M}$ through q. Since every point different from q of G lies on a line that meets N and misses N_2, we have

$$\sum_{p \in N} (r_p - n - 1) \geq k_G - 1.$$

Since the left-hand side is just $|\mathcal{X}|$, we obtain using equation (1)

$$v - n + k_G - 1 \leq \sum_{M \in \mathcal{M}} k_M = k_G + (|\mathcal{M}| - 1)n = k_G + (n - 1)n,$$

a contradiction. Thus statement (a) is proved.

Next we show

(b) An n-line of $\mathcal{M}$ meets no other line of $\mathcal{M}$.

Assume to the contrary that an n-line N' meets another line of $\mathcal{M}$ in a point p. Then p is not on L, since every point of L has degree $n + 1$. Since $1 + x_p$ is the number of lines of $\mathcal{M}$ through p, we have $x_p \geqq 1$. Therefore p lies on a line X of $\mathcal{X}$. Then X meets N and N', contradicting part (a).

We define the number z by $b = n^2 + n + 1 + z$. Then $v \geqq b - n = n^2 + 1 + z$, and L is disjoint to exactly z lines. It follows from inequality (2) that $\mathcal{M}$ has $s \geqq 1 + z$ lines $N_1, \ldots, N_s$ of degree n. By part (b), these lines are mutually disjoint.

Assume that $\mathcal{X}$ contains a line X. By part (a), X is disjoint to every line N_i. Consider a point $q \neq X \cap N$ of $\mathcal{X}$. Then $x_q \geqq 1$, and so q lies on (at least) two lines G_1 and G_2 of $\mathcal{M}$. By part (b), the lines G_1 and G_2 miss every line N_i. Since q lies on $2 > k_L - n$ lines that meet L and miss N_i, there is a line H_i through q which meets N_i and misses L, $i = 1, \ldots, s$. By part (a), $H_i \neq H_j$ for $i \neq j$. Hence L is disjoint to at least s lines, contradicting $s \geqq 1 + z$.

Consequently, $\mathcal{X} = \phi$, that is $x_p = 0$ for every point p outside N. This implies that $\Pi = \mathcal{M} \cup \{N\}$ is a parallel class of S. We have $|\Pi| = n + 1$. By adjoining Π as an infinite point to S we get the linear space S'. By the choice of S, the linear space S' has a line G of degree $n + 1$ that has a point q of degree at least $n + 2$. The line G must be an n-line of Π, and q must be a point of S (since $|\Pi| = n + 1$). However, since $\mathcal{X} = \phi$, the line N has only points of degree $n + 1$, and we can show the same for every n-line that meets L. This contradiction completes the proof of the proposition. □

For a line L, recall that m_L is the number of lines that miss L. We also define $t_L = \Sigma(r_p - n - 1)$, where the sum runs over all points p of L. Notice that a long line L satisfies $b = n^2 + n + 1 + t_L + m_L$.

Proposition 6.2.2. Suppose that G and L are long lines and that the line H meets G and misses L. If G and L meet in the point p, then $t_L \geqq (k_H - 1) + (r_p - n - 1)$.

PROOF. Since G and L have the same degree, every point different from $H \cap G$ of H lies on a line which meets L and misses G. Since L meets $t_L - (r_p - n - 1)$ lines that miss G, the assertion follows. □

Proposition 6.2.3. If Y is a line of maximal degree that misses a long line L, then Y is not hyperideal.

PROOF. Assume to the contrary that Y is hyperideal and denote by $n + 1 + a$ the minimum point degree of Y. Since Y misses L, we have $a \geqq 1$. Put k equal k_Y. We derive a contradiction in several steps.

Step 1. $v \leqq n^2 + a(k - 1)$.

This follows by counting the points from a point p of degree $n + 1 + a$ of Y. Since Y is hyperideal, the $n + 1$ lines joining p to a point of L have degree at most n. By the choice of Y, the lines through p that miss L have degree at most k. Hence $v \leqq 1 + (n + 1)(n - 1) + a(k - 1)$.

Step 2. Every long line G satisfies $t_G \leqq k - a - 2$.

Since Y is hyperideal, it misses G. Hence Y meets at least $1 + k(a - 1)$ lines that miss G; that is $m_G \geqq 1 + k(a - 1)$. Since by step 1, $b \leqq v + n \leqq n^2 + n + a(k - 1)$ and $b = n^2 + n + 1 + t_G + m_G$, the assertion follows.

Step 3. If the line X is ideal but not hyperideal, then $k_X \leqq k - a - 1$.

Since X is ideal, there is a long line L_1 that misses X, and since X is not hyperideal, there is a long line L_2 that meets X. By Proposition 6.2.2, $t_{L_1} \geqq k_X - 1$. The assertion follows from step 2.

Step 4. Every ideal line is hyperideal.

Let $\mathcal{N}$ be the set consisting of all ideal lines that are not hyperideal and assume that $\mathcal{N} \neq \phi$. Choose a line T of maximum degree of $\mathcal{N}$. Then there is a long line L_1 that meets T in a point u, and a long line L_2 that misses T. Choose a point $w \in T - u$ of minimum degree. Put $r_w = n + 1 + e$; denote by g the number of real lines through w, and by h the number of lines of $\mathcal{N}$ through w. Since $T \in \mathcal{N}$, we have $g \leqq k_{L_1} - 1 = n$. Since T meets L_1 and misses L_2, there exists a line R (of $\mathcal{N}$) through w that meets L_2 and misses L_1. Proposition 6.2.1 implies $t_{L_1} \geqq k_R - 1$. Using $m_{L_1} \geqq (k_T - 1)e$, we obtain

$$b \geqq n^2 + n + 1 + (k_R - 1) + (k_T - 1)e. \tag{3}$$

Let c be the number of long lines through w and assume that $c \geqq 1$. Then no line through w is hyperideal, that is $r_w = g + h$. Hence R and the e lines through w that miss L_2 have degree at most k_T. Thus

$$v \leqq k_R + e(k_T - 1) + cn + (n - c)(n - 1).$$

In view of $c \leqq n$ and $b - v \leqq n$, we see that inequality (3) implies $n = c$. Since the n long lines through w miss the hyperideal line Y, we have $k + n \leqq r_w = n + 1 + e$. That is, $e \geqq k - 1$. For a long line G through w, we have by step 2

$$k - a - 2 \geqq t_G \geqq r_w - n - 1 = e \geqq k - 1,$$

a contradiction.

We have shown that w does not lie on long lines. Let p be a point of degree $n + 1$ of L_1 (see Proposition 6.1.7(b)) and denote by d the number of long lines through p. Since w lies on at most n real lines, every long line through p meets a line of $\mathcal{N}$ through w. It follows that the lines of $\mathcal{N}$ through w together have at least d points, so that $d \leqq 1 + h(k_T - 1)$ and hence

$$v \leqq 1 + r_p(n - 1) + 1 + h(k_T - 1) = n^2 + 1 + h(k_T - 1).$$

Comparing this with inequality (3) and $b - v \leqq n$, we see that $h(k_T - 1)$

$\geqq e(k_T - 1) + k_R - 1$, and so $h \geqq e + 1$. Since w does not lie on long lines, it follows that

$$\begin{aligned} v - 1 &\leqq g(n-1) + k_R - 1 + (h-1)(k_T - 1) + (r_w - g - h)(k-1) \\ &= g(n-k) + k_R - 1 + (h-1)(k_T - k) + (n + e - 1)(k-1) \\ &\leqq g(n-k) + k_R - 1 + e(k_T - k) + (n + e - 1)(k-1) \\ &\leqq g(n-k) + k_R - 1 + e(k_T - 1) + (n-1)(k-1). \end{aligned}$$

Comparing with inequality (3), and using $g \leq n$, we conclude that $k + n \leq 1$, a contradiction. So $\mathcal{N} = \phi$.

Now we can complete the proof of the proposition. By Proposition 6.2.1, there is a long line G with a point p of degree at least $n + 2$. By Proposition 6.1.7(b), G has also a point q of degree $n + 1$. Since $v \geqq n^2 + 2$, the point q lies on a second long line G'. Now, every line through p that misses G' is ideal, but it is not hyperideal, since it meets G. This contradicts the conclusion of step 4.

Now we can complete the proof of Totten's theorem. Consider an ideal line X of maximum degree. In view of Proposition 6.2.3 it meets a line L of degree $n + 1$. Put $p_0 = X \cap L$ and choose a point q of minimum degree of $X - p_0$. We denote the degree of q by $n + 1 + a$. Since X is ideal, we have $a \geqq 1$. Every point different from p_0 of X lies on a line missing L; therefore $m_L \geqq (k_X - 1)a$. Let T be a line of maximum degree missing L. By Proposition 6.2.3 there exists a line L' of degree $n + 1$ that meets T. By Proposition 6.2.2 we have $t_L \geqq k_T - 1$. It follows that

$$b \geqq n^2 + n + 1 + t_L + (k_X - 1)a \geqq n^2 + n + 1 + k_T - 1 + (k_X - 1)a.$$

Now consider the lines through q to obtain an upper bound for the number of points. The lines $L_1, \ldots, L_n \neq X$ which join q to a point other than p_0 of L have degree at most $n + 1$, and the a lines through q that miss L have degree at most k_T. Hence

$$v \leqq 1 + n \cdot n + k_X - 1 + a(k_T - 1).$$

It follows that $b - v \geqq n + (a - 1)(k_X - k_T)$. Since X is an ideal line of maximum degree, we have $k_X \geqq k_T$. The facts $a \geqq 1$ and $b - v \leqq n$ imply that equality holds in all the above inequalities. In particular, $b - v = n$, the lines L_i have degree $n + 1$, $t_L = k_T - 1$, and the lines through q that miss L have degree k_T. In view of $t_L = k_T - 1$, Proposition 6.2.1 shows that the point $L \cap L'$ has degree $n + 1$. Since each of the lines through q that misses L could play the role of T and since therefore each of the lines L_i could play the role of L', we see that the points $L \cap L_i$

have degree $n + 1$, that is, the n points different from p_0 of L have degree $n + 1$.

In the same way we can show that each point different from q on one of the lines L_i has degree $n + 1$. Thus, the real points are the n^2 points of $(\cup L_i) - q$. Since every real line meets every line L_i, we see that each real line has exactly n real points, which implies that two ideal points are joined by an ideal line (since every (real) line has degree at most $n + 1$). Hence the real points and lines form an affine plane A of order n, and the ideal points and lines form some linear space S^*. Since $b - v = n$ and since A has n more lines than points, we see that S^* has the same number of points and lines, that is S^* is a generalized projective plane. Therefore the linear space S is an affine plane with a generalized projective plane at infinity. □

6.3 The algebraic approach

Now we shall present a second method to prove the theorem of J. Totten. The idea goes back to J. C. Fowler (1984), but we use a modified proof following K. Metsch (1991).

Proposition 6.3.1. Let q be a point of a linear space S and consider a set $\mathcal{M}$ of t lines of S through q. Suppose that each line $L \in \mathcal{M}$ has the following property: Each point other than q of L has degree at least k_L and some point of L has degree at least $k_L + 1$. Then $b \geqq v + t - 1$.

PROOF. We denote the lines of $\mathcal{M}$ by $L_1, \ldots, L_t$. Let $p_1, \ldots, p_w$ be the points that lie on no line of $\mathcal{M}$, and let $p_{w+1}, \ldots, p_{v-1}$ be the points not equal q of the lines of $\mathcal{M}$. For $i \leqq w$, define $R_i = \{i\}$ and for $i = 1, \ldots, t$ define $R_{w+i} = \{j | p_j \in L_i\}$. Then the set $\{R_i | i = 1, \ldots, a = w + t\}$ forms a partition of the set $R = \{1, \ldots, v - 1\}$. We define the $(v - 1) \times (v - 1)$-matrix A as follows:

$$
\begin{array}{ll}
a_{ii} = r_{pi} & \text{for } i \leqq w, \\
a_{ii} = r_{p_i} - 1 & \text{for } i = w + 1, \ldots, v - 1, \\
a_{ij} = 0 & \text{if } i \text{ and } j \text{ lie in the same set } R_k, \\
a_{ij} = 1 & \text{if } i \text{ and } j \text{ do not lie in the same set } R_k.
\end{array}
$$

Let S' be the incidence structure obtained from S by removing the point q and the lines of M. If C is the incidence matrix corresponding to the order $p_1, \ldots, p_{v-1}$ of points, then $A = CC'$. The crucial point of the proof is to show that A is invertible. Since C is a $(v-1) \times (b-t)$-matrix, it then follows that $v - 1 \geqq b - t$.

Suppose that $A(x_i) = 0$, where (x_i) is a $(v-1)$-vector of indeterminates x_i. We define

$$s_i = \frac{1}{r_{p_i}} \text{ for } i \leq w \quad \text{and} \quad s_i = \sum_{j \in R_i} \frac{1}{r_{p_j} - 1} \quad \text{for } i = w+1, \ldots, a.$$

We have $0 < s_i < 1$ for $i \leqq w$, since every point has degree at least 2. For $i > w$, our hypothesis yields that $0 < s_i < 1$.

We define the numbers

$$y_j = \sum_{i=1}^{v-1} x_i - \sum_{i \in R_j} x_i$$

and the $a \times a$-matrix $B = (b_{ij})$ with $b_{ii} = 1$ and $b_{ij} = s_j$ for $i \neq j$. In view of $A(x_i) = 0$ we have

$$a_{ii}x_i + y_j = 0 \qquad \text{for } j = 1, \ldots, a \qquad \text{and} \qquad i \in R_j. \tag{4}$$

It follows that

$$y_j = \sum_{\substack{k=1 \\ h \neq k}}^{a} \sum_{i \in R_k} x_i = \sum_{\substack{k=1 \\ h \neq k}}^{a} \sum_{i \in R_k} \frac{-y_k}{a_{ii}} = -\sum_{\substack{k=1 \\ h \neq k}}^{a} y_k s_k,$$

and so $B(y_j) = 0$. Since $0 < s_i < 1$ for $i = 1, \ldots, a$, it is easily seen that B is invertible. It follows that $(y_j) = 0$, so that also $(x_i) = 0$ by equation (4). Hence the matrix A is invertible. □

Proposition 6.3.2. The linear space S is an affine plane of order n with a generalized projective plane at infinity.

PROOF. We shall call the points of degree $n + 1$ and the lines with a point of degree $n + 1$ *real*. The other points and lines will be called *ideal*. Also we call a line *good* if it has degree $n + 1$ and if it has a unique ideal point. Let v_0 (or v') be the number of real (or ideal) points and let b_0 (or b') be the number of real (or ideal) lines. We prove the proposition in five steps.

Step 1. Every ideal point lies on a good line.

Let q be an ideal point and assume that no line through q is good; that is, every line through q has either degree at most n or a second ideal point. Then the preceding proposition shows that $b \geqq v + r_q - 1$, a contradiction.

Step 2. Every real line contains at most one ideal point.

Assume to the contrary that there is a real line R that has two ideal points q_1 and q_2. By the preceding step there exist good lines L_1 and L_2 containing q_1 and q_2 respectively (see Figure 6.3.1). The point q_2 lies on a line $L_1{}^*$ missing L_1. Let q be a second point of $L_1{}^*$. Since $L_1{}^*$ misses L_1, the point q is ideal. By step 1 it lies on a good line L. Since L and L_1 have the same degree and since q_2 lies on the line $L_1{}^*$, which meets L and misses L_1, the point q_2 lies on a line M that meets L_1 and misses L. The line M and hence every point of M is ideal, since it misses L. Since q_1 is the only ideal point of L_1, we see that M and L_1 must meet in q_1. But then $M = q_1q_2 = R$, a contradiction, since R is real.

Step 3. We have $b_0 = n^2 + n$ and every ideal point lies on exactly n real lines. Also $v_0 \leqq n^2$ with equality if and only if every real line has exactly n real points.

Let q be an ideal point, L a good line through q, and p a real point of L. Since $v \geqq n^2 + 2$, the point p is contained in a second line X of degree $n + 1$ (see Figure 6.3.2). Let L'' be any line through q that misses X, and let q' be a second point of L''. Since X is parallel to L'', the line L'' and the point q' are ideal. Let L' be a good line passing through q'. For every real point p' of L' the line qp' is real. On the other hand, every real line has a point of degree $n + 1$ and therefore meets L'. Since L' is a good line, every point not equal q' of L' is real; therefore q lies on exactly n real lines. Since every other point of L is contained only in real lines and since L meets every real line, it follows that there exist exactly $n^2 + n$ real lines. This implies that every real line has at most n real points (a real line with $n + 1$ real points would meet $n^2 + n$ other real

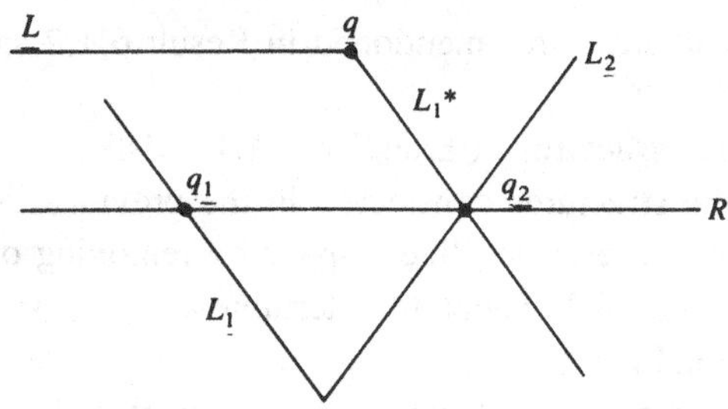

Figure 6.3.1.

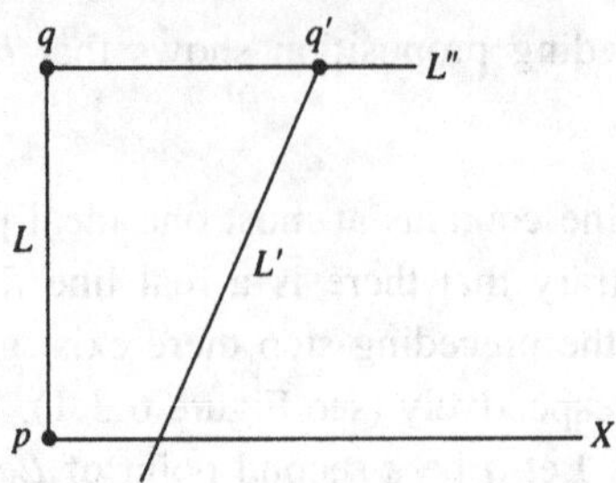

Figure 6.3.2.

lines). Counting the number of real points from any real point x shows therefore that $v_0 \leqq n^2$ with equality if and only if every line through x has exactly n real points.

Step 4. The linear space S is an affine plane with a generalized projective plane at infinity.

In view of $b_0 = n^2 + n$, there are $b - n^2 - n \geqq 2$ ideal lines. By definition, an ideal line contains only ideal points. Since lines joining two ideal points are ideal by step 2, it follows that the ideal points and lines form a linear space P. It has v' points and b' lines. The Fundamental Theorem shows that $b' \geqq v'$. In view of

$$n^2 + n + b' = b_0 + b' = b \leqq v + n = v_0 + v' + n \leqq n^2 + n + v',$$

we conclude that $b' = v'$ and $v_0 = n^2$. Step 3 shows that every real line has exactly n real points. Hence, the $n^2 + n$ real lines and the n^2 real points form an affine plane of order n. Therefore S is an affine plane of order n with the linear space P at infinity. Since $b' = v'$, the Fundamental Theorem shows that P is a generalized projective plane. □

6.4 Exercises

1. Check that all structures mentioned in Result 6.1.2 satisfy $(b - v)^2 \leqq b$.
2. Which of the structures of Exercise 6.4.1 satisfy $v < (b - v)^2 \leqq b$?
3. Which of the structures considered in the previous exercise can be obtained from a restricted linear space by removing one point?
4. Use the theorem of J. Totten to determine all linear spaces satisfying $b = v + 1$ and $b = v + 2$.
5. Show that there does not exist a linear space satisfying $v = n^2 + n + 1$ and $n^2 + n + 1 < b < n^2 + 2n + 1$.

6. Check that the (3,4)-cross is the only restricted linear space whose number of lines can be written in the form $b = n^2 + n + 2$.
7. Let n be an integer of the form $n = m^2 + m$, $m \in N$. Show that there is no linear space S on $v = n^2 + n + 1$ points with

$$n^2 + 2n + 1 < b < n^2 + 2n + m + 1,$$

which is constructed by taking an affine plane of order n and adjoining a linear space at infinity.

6.5 Research problems

1. Given an integer $n \geqq 2$, what is the smallest integer $v_0 = v_0(n)$ such that every $(n + 1, 1)$-design with $b \leqq n^2 + n + 1$ lines and $v \geqq v_0$ points can be embedded in a projective plane? (See Metsch 1991, p. 60.)

6.6 References

Fowler, J. C. (1984), A short proof of Totten's classification of restricted linear spaces. *Geom. Dedicata* 15, 413–422.

Metsch, K. (1991), *Linear spaces with few lines*. Springer-Verlag Lecture Notes in Mathematics 1490, Berlin, Heidelberg, New York, London.

Stanton, R. G. and Kalbfleisch, J. G. (1972), The λ-μ problem: $\lambda = 1$ and $\mu = 3$. Proc. Second Chapel Hill Conf. on Combinatorics, Chapel Hill, North Carolina, 451–462.

Totten, J. (1975), Basic properties of restricted linear spaces. *Discrete Math.* 13, 67–74.

Totten, J. (1976a), Classification of restricted linear spaces. *Can. J. Math.* 28, 321–333.

Totten, J. (1976b), Parallelism in a restricted linear space. *Discrete Math.* 14, 395–398.

7

d-Dimensional linear spaces

7.1 The definition

Given an arbitrary linear space $S = (p, \mathcal{L})$, there is a natural way of introducing further structure on S by distinguishing subsets of p with the property that any element of $\mathcal{L}$ containing at least two points of such a subset must be wholly contained within the subset. We call any subset of p with the above property, along with the induced lines, a *subspace* of S. Clearly ϕ and S themselves are subspaces. Moreover, the intersection of any set of subspaces is again a subspace (the empty intersection being S), and so a closure space is induced on S, the closure of any set X of points of p being the intersection of all subspaces containing X.

There are various ways in which a 'dimension' can now be assigned to subspaces. In any case, we normally wish to assign dimensions 0 and 1 to the points and lines respectively, and require that $V \subset W$ implies that the dimension of V be less than or equal to the dimension of W. We shall make the following definition.

If it is possible to choose *a subset* of the set of subspaces, including all the points and lines, and such that the intersection property is retained, and in addition assign to each element of this subset a unique integer between 0 and d for some fixed integer $d \geq 2$ so that

(i) each point is assigned 0 and each line is assigned 1;
(ii) if V is assigned i, for any point p not in V there is a unique subspace W on V and p which is assigned $i + 1$;
(iii) S is assigned d,

then we say that S is a *d-dimensional linear space*.

In the above, it is not necessary to assume that d is finite, but if d is infinite we normally add the condition that $V \subset W$ implies $i < j$ where

V and W have been assigned i and j respectively. The reader is invited to prove that in the finite case, this follows automatically.

If a subspace is assigned the number i, we refer to it as an *i-subspace* or simply as an *i-space*.

Note that any linear space can be assigned the structure of a 2-dimensional linear space.

If $X_1, \ldots, X_s$ are sets of points of a d-dimensional linear space, then $\langle X_1, \ldots, X_s \rangle$ denotes the intersection of all subspaces on $X_1, \ldots$ and X_s. It is easy to see that this is the *smallest* subspace on the X_is.

We generalize the notation v, b and r_p as follows. If V is an i-space, v_V or v_i and b_V or b_i will denote respectively the number of points and lines of V; r_p^i denotes the number of lines on p in V.

7.2 Problems in *d*-dimensions

The problems we have considered in earlier chapters can be generalized to the d-dimensional case without much difficulty. For instance, suppose S is a d-dimensional linear space in which $b - v = n^4 + n^2$. Is S a projective 3-space? (Note that in projective 3-space, $b = n^4 + n^3 + 2n^2 + n + 1$ and $v = n^3 + n^2 + n + 1$.) What can you say if $b - v \leq n^4 + n^2$, with some lower bound on $b - v$; or if $b - v \geq n^4 + n^2$, with some upper bound? These questions generalize the restricted linear space types of problems, and have been considered by Hafner (1993) and Metsch (1993). The 3-dimensional case turns out to be at least as difficult as the planar case.

Let S be a linear space in which each point lies on $n^2 + n + 1$ lines and such that given any line L not on the point p there are precisely n^2 lines on p missing L. Is S a 3-dimensional projective space? This question and similar questions generalize the semi-affine problems which were considered in Chapters 4 and 5.

If a d-dimensional linear space has constant line size, what can be said? Or if the line sizes are two consecutive numbers, is it possible to deduce something? These problems are too unwieldy without additional constraints, but we might assume in addition, for instance, that there is an upper bound on the number of points in each 2-subspace.

The complementation problem is one which has not been addressed very much in dimensions higher than 2. But many obvious questions come to mind as generalizations of the problems of Chapter 3. We pose such questions in the research exercises at the end of the chapter.

In fact, not much work has been done at all on problems of the types mentioned so far. For the most part, research on higher-dimensional linear spaces has been concerned with the existence of embedding in projective spaces of the same dimension. In case a linear space of dimension at least 3 has at each point the structure of an affine space or of a projective space (an idea which we shall make more precise in a later section of this chapter), a great deal is known. We present the results later.

In the next section, we wish to concentrate on questions concerning linear spaces in which all 2-dimensional subspaces are essentially the same. Such questions can be viewed in a general way as 'local' generalizations of earlier chapters. In Section 7.5 we consider 'global' generalizations of the earlier chapters; that is, we apply conditions to the whole space, rather than just to a particular set of i-spaces.

For the remainder of the chapter, the word *plane* will always denote a 2-subspace of a linear space.

7.3 π-Spaces

We consider in this section those d-dimensional linear spaces, $d \geq 3$, in which each plane is isomorphic to one of a given set of planes. If the set contains a single plane π, then the space is called a *π-space*. If the set contains many planes, the possibilities for S are too numerous and the problem becomes less interesting. For the most part, we shall be concerned with π-spaces only.

The first result is easy. It follows immediately from the definitions. Suppose that the set of planes of S is precisely the set of subspaces generated (using linear closure) by three non-collinear points. Then *S is a projective space precisely when each of its planes is projective*; moreover, if S has dimension at least 3, then all of its planes are Desarguesian and of the same order.

The second result is not quite so trivial. It is the comparable result for affine spaces, and was already mentioned in Section 1.3. We once again assume that the planes are precisely the subspaces generated by three non-collinear points. The reader is referred to Chapter 1 to review the definition of affine space.

Theorem 7.3.1. (Buekenhout 1969.) Let S be a non-trivial linear space such that each plane is affine and such that each line has at least four points. Then S is an affine space.

PROOF. By Section 1.2 it suffices to show that one can define an equivalence relation $\|$ on the lines of S such that (i) for each point p and line

L there is a unique line L' such that $p \in L'$ and $L \| L'$, (ii) if p, q, r, s, t are points such that $pq \| rs$ and $t \in pr$, then either $t \in qs$ or the lines pq and ts have a common point, (iii) there exist two disjoint lines L and L' such that $L \nparallel L'$.

We define $\|$ in the most natural way: $L \| L'$ precisely if $L = L'$ or if L and L' are coplanar and disjoint. With this definition, (i) and (ii) are easily seen to follow, and (iii) fails only if S is itself an affine plane. Thus it remains only to show that $\|$ is an equivalence relation. Since the reflexive and symmetric properties are immediate, and transitivity is trivial if lines L, L' and L'' are not all distinct or if they are coplanar, we need only show that if L, L' and L'' are distinct and non-coplanar, $L \| L'$ and $L' \| L''$, then $L \| L''$.

As a first, and major step in the proof, we prove the following: let π be any plane and L any line meeting π in a single point p; let X be the set of all points lying in any plane on L which meets π in a line; then X is a subspace.

We need only prove that if x and y are any two elements of X, then each point of the line xy is in X. If xy and L are coplanar, this is trivial; so we may suppose that they are not. Let x' and y' be respectively the points of intersection with L of the lines on x and y parallel to the line of π in the planes determined by x and L and by y and L. By assumption, L has a point $q \neq p, x', y'$. Then xq and yq meet π in points x'' and y'' respectively. If $xy \| x''y''$, then for any point z of xy, the line zq meets $x''y''$ in a point of π, and so $z \in X$. So let $xy \cap x''y'' = r$. (See Figure 7.3.1.) Then there is precisely one point z of L for which the line zq misses $x''y''$. It suffices to show that $z \in X$.

We consider first, then, the case $x' = y'$. Let u be any point of xy except

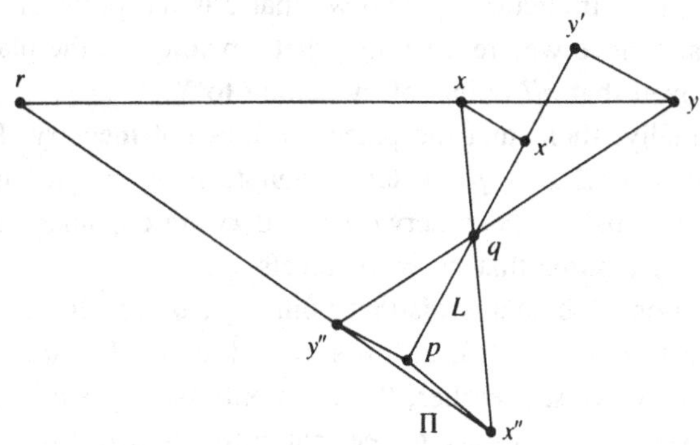

Figure 7.3.1.

z or r. Define u' as for x' and y'. If $u' \neq x'$, then the line $x'u$ meets π in a point different from r. This point generates a plane together with r and with u which contains x and y as well as $x' = y'$. It follows that one of the lines xx', yy' meets π, contrary to their definition. Hence $u' = x'$ for all $u \in xy - \{z, r\}$.

It is now easy to see that the plane on x, y and x' cannot meet π in a line. Let v be a fixed point of $xx' - \{x, x'\}$. Then $v \in X$ by definition. Consider the line vy. If this line is not entirely contained in X then we can argue as above for xy to obtain the result $vy \cap \pi \neq \phi$. But then $r \in vy$, a contradiction. Thus $vy \subset X$. By assumption, there is a point $w \neq v, x, x'$ of xx', and as for vy, we see that $wy \subset X$. For any point $u \in vy$ or wy, the line $ux' \subset X$. But every line on x' meets either vy or wy. Hence the whole plane on x, y and x' is in X. In particular, $L = xy \subset X$.

We may henceforth suppose that for all points u and v of $xy - \{z, r\}$, $u \neq v$ implies $u' \neq v'$. Thus, for *each* point $q \in L - p$, it is possible to find a point $u \in xy - r$ such that $u \in X$ and $u' \neq q$. Defining u'' as earlier, we see that $u'' \neq r$. The line through q and parallel to xy meets $x''y''$ in a point s. Let M be any line of π passing through p. If M meets $x''y''$ in a point $t \neq s$, then ts meets xy and so the plane on L and M meets xy in precisely one point. We consider the only two possible exceptions to this, corresponding respectively to $M = ps$ and $M \| x''y''$. Let π_1 be the plane on L and s, and π_2 the plane on L and on the line through p parallel to $x''y''$. π_1 cannot meet xy, as such a point of intersection would have to be in qs, which is parallel to xy. If π_2 meets xy in a point x, we shall show that z is the z of the first part of this argument – that is, that the line zq is parallel to $x''y''$. Suppose this is not the case, and that $zq \cap x''y'' = z''$. Then z'' belongs to the plane on L and M also, which implies that $x''y''$ is in this same plane. But this forces xy also to be coplanar with L, and gives us a contradiction. It follows that z is the point of our earlier discussion; and since we are assuming that z belongs to the plane π_2, we have just proved that *all* points of xy belong to X.

Suppose finally, then, that the plane π_2 does not meet xy. In fact, π_2 is a function of $q \in L - p$, so let us denote it by $\pi_2(q)$. Then all the planes $\pi_2(q)$ are distinct (as otherwise L and xy are coplanar), and by the above, we may assume that no $\pi_2(q)$ meets xy.

By assumption, L contains distinct points q_1, q_2, q_3 different from p. The planes $\pi_2(q_i)$, $1 \leq i \leq 3$, all miss xy and are on L and some line M of π on p. But as we saw earlier, there are only two possible such planes which miss xy. Now it is easy to see that $\pi_2(q_i) = \pi_2(q_j)$ for $i \neq j$ leads to the contradiction L and xy coplanar, and hence we have reached a final contradiction.

In conclusion, the set X is a subspace.

We now return to the situation of three distinct lines L, L', L'' such that $L \| L'$ and $L' \| L''$, and the three lines are not all contained in a common plane. It suffices to prove that L and L'' are coplanar. Let π be the plane on L and L', let $p' \in L'$ and $p'' \in L''$. Let X be the set of points constructed as described above from the plane π and the line $p'p''$. Let M be the line on p'' parallel to L. Let N be the line on p' parallel to M. The set X contains each point of L and also the point p''. Since, by our first argument, it is a subspace, it also contains each point of M. Thus the plane $\langle M,p' \rangle$ meets π in a line. If M itself meets π, then $\langle L,L' \rangle = \langle L,M \rangle$, a contradiction. Thus M is parallel to the line of intersection of π and $\langle M,p' \rangle$. This line is therefore N. Thus $N \subset \pi$, and if $N \neq L'$, it meets L in a point. But this forces $\langle L,M \rangle = \langle N,M \rangle = \pi$, again a contradiction. Thus $N = L'$. But now $L' \| L''$ forces $M = L''$ and we are done. □

An alternative proof can be found in Karzel and Pieper (1970); see also Sörensen (1988).

Consider the statement of Theorem 7.3.1 without the restriction that each line have at least four points. Without this restriction, the same result cannot be obtained. In fact, this was already known in 1960, a number of years before Buekenhout's paper was published.

M. Hall, Jr. (1960) constructs examples of systems in which each triangle is in a unique affine plane of order 3, but which are not themselves affine spaces. We leave it to the reader both to try to find independently such examples, and to study Hall's paper.

The next major result we propose to present here is a result of L. Teirlinck (1975) in which he considers a very nice generalization of the above results for projective and affine spaces: Suppose each plane of the space is affine or projective; what can be said?

Theorem 7.3.2. If every plane of a linear space S is projective or affine, then all planes are projective or all planes are affine.

PROOF. We prove first that all lines have the same cardinality, which we shall then call k. So let L and L' be lines of S. If they meet, then $\langle L,L' \rangle$ is a plane, and hence either projective or affine. So $k_L = k_{L'}$. If they are disjoint, take $p \in L$, $p' \in L'$. Using the above argument, we obtain $k_L = k_{pp'} = k_{L'}$.

Now suppose that S has both an affine plane A and a projective plane π. We claim that some affine and projective plane pair have a common line. Take a line L of A and a line $L' \neq L$ of π. Take $p \in L - L'$,

$p' \in L' - L$. If $\langle L,p' \rangle$ is projective, use L. If $\langle L',p \rangle$ is affine, use L'. If $\langle L,p' \rangle$ is affine and $\langle L',p \rangle$ projective, use pp'.

In any event, we now use L to indicate a line common to the affine plane A and projective plane π. Also, let x be a fixed point of $A - L$.

Case 1. $k \geq 4$.

Let L' be a line of A parallel to L, $L' \neq L$, and let $x \notin L'$. For any $y \in \pi - L$, note that $\langle L',y \rangle \cap \pi = \{y\}$, using the fact that π is projective.

Now let $p' \in L'$, $p = xp' \cap L$, and consider the line $p'y$ in the plane $\langle x,p,y \rangle$ for fixed $y \in \pi - L$. Since $k \geq 4$, there is a point $u' \in p'y$ such that $u' \neq p', y$, and the line xu' intersects the line py in some point u.

Take $q' \in L' - p'$, $q = L \cap xq'$. The plane $\langle x,q',u' \rangle$ is an affine plane in which $q'u' \| qu$, otherwise $q'u'$ meets qu in a point distinct from y which belongs to $\langle L',y \rangle$ since $q'u'$ is in $\langle L',y \rangle$, contradicting $\langle L',y \rangle \cap \pi = \{y\}$.

Let $z' \in q'u'$, $z' \neq u'$, and $z = xz' \cap qu$, again using $k \geq 4$. Then $p'z' \cap \pi = \phi$ and $p'z' \| pz$ in the plane $\langle x,p,z \rangle$.

Now consider the plane $\langle M,x \rangle$, where M is any line of π with $p \in M$, $u \notin M$. Let $M \cap qu = w$ and $xw \cap q'u' = w'$. In the plane $\langle x,p,w \rangle = \langle x,M \rangle$, $p'w' \| pw = M$, and $p'w' \subset \langle L',y \rangle$.

Therefore, for all lines M of π with $p \in M$, except perhaps one, the plane $\langle x,M \rangle$ is affine; and in $\langle x,M \rangle$ the line on p' parallel to M lies in $\langle L',y \rangle$.

Now take a second point $y' \in \pi - L$, $y' \neq y$. The conclusions above also hold for y'. But $\langle L',y \rangle$ and $\langle L',y' \rangle$ then have at least 4–2 lines in common, implying they are the same plane. This gives a final contradiction.

Case 2. $k = 3$.

In this case, we introduce the following notation. For $p \neq q$ two points of S, $p + q$ denotes the third point of the line pq. If X is a subset of the point set of S and if $p \notin X$, then $p + X = \{p + x | x \in X\}$. We remark that $p + L$ is a line if and only if $\langle p,L \rangle$ is affine.

Since $\langle x,L \rangle = A$ is affine, $x + L$ is a line of $x + \pi$.

For $z \in (x + \pi) - (x + L)$, then $|\langle x + L, z \rangle \cap \pi| \leq 1$ because if $\langle x + L, z \rangle \cap \pi$ were a line, this line would meet L in some point, forcing the contradiction $\langle x + L, z \rangle = A$. On the other hand, $z + (x + L)$ is a subset of $\langle z, x + L \rangle$ having at least two points in $x + \pi$; hence $|\langle x + L, z \rangle \cap (x + \pi)| \geq 6$. If there is a point $p \in (x + \pi) - \langle x + L, z \rangle$, the fact that $|\langle x + L, p \rangle \cap (x + \pi)| \geq 6$ implies $|\langle x + L, p \rangle \cap \langle x + L, z \rangle| \geq 5$. Hence $\langle x + L, p \rangle = \langle x + L, z \rangle$ giving the contradiction $p \in \langle x + L, z \rangle$. So $x + \pi \subset \langle x + L, z \rangle$, and consequently $\langle x + L, z \rangle \cap \pi = \phi$.

Now for distinct points $u = x + p_1$ and $w = x + p_2$ of $x + \pi$, either

$\langle x, p_1, p_2\rangle$ is affine and $u + w \in x + \pi$, or it is projective and $u + w \in \pi$. It follows now that $x + \pi$ is a subspace, and so $x + \pi = \langle x + L, z\rangle$.

If we take a line $M \subset \pi$, then $x + M \subset \langle x, M\rangle \cap (x + \pi)$ is a line because $|\langle x,M\rangle \cap (x + \pi)| \leq 3$. Therefore $\langle x,M\rangle$ is affine by the remark at the beginning of this proof.

Take a point $q \in \pi - L$, and a line M' so that $q \in M' \subset \pi$. Then $\langle x,M'\rangle$ is affine, $x + \pi$ is projective, and the line $x + M'$ is common to both planes. Also, $q \in \langle x,M'\rangle$, and $x + L$ is a line of $x + \pi$. It follows that $q + (x + L)$ is a line.

For $y \in L$, $\langle xy,q\rangle = \langle x,qy\rangle$ is affine, and so $q + xy$ is a line by the remark above. This line has the points $(x + y) + q$ and $x + q$ in common with $\langle x + q, q + (x + L)\rangle$. So $q + xy \subset \langle x + q, q + (x + L)\rangle$ and $q + y \in \langle x + q, q + (x + L)\rangle$. Thus $\langle x + q, q + (x + L)\rangle$ contains three distinct points of π, no one of which lies on L. Since π is projective, this implies that these points are non-collinear, so $\langle x + q, q + (x + L)\rangle = \pi$, a final contradiction. □

The next result specializes the situation somewhat. We suppose not only that each plane is isomorphic to a fixed linear space π_0, but also that the structure of the subspaces through each point of π is isomorphic to π_0. In this case, A. Delandtsheer (1983) proves that S is either a $PG(3,n)$ or a 3-dimensional generalized projective space consisting of two disjoint lines of the same size, all other lines having size 2.

Theorem 7.3.3. (Delandtsheer 1983.) Let S be a 3-dimensional finite linear space in which every plane is isomorphic to the linear space π_0, and in which the residue at every point is isomorphic to π_0. Then π_0 is a (possibly degenerate) projective space PG(3,n) or a linear space consisting of two disjoint lines of the same size, all other lines having size 2.

PROOF. We claim first that the set of line degrees equals the set of point degrees in π_0. To see this, fix a point p of S and a plane π on p. Then the number of lines on p in $\pi \approx \pi_0$ is the number of points on the 'line' π in S_p, the structure of lines and planes of S on p. Conversely, if any 'line' π_0 in S_p, for any point p of π_0, contains k 'points', then p is on k lines in π.

We claim, secondly, that the degree of any line L in S is the same as the degree of the point L in S_p where p is any point of L. Letting $k = |L|$ and $f(k) = \dfrac{v - k}{v_0 - k}$, the number of planes on L, where v and v_0 are respectively the number of points of S and of π_0, we see that the result of the last paragraph allows us to conclude that the sets $\{k\}$ and $\{f(k)\}$ are identical. But the function $k \to f(k)$ is increasing (that is, $k_1 \leq k_2$ implies $f(k_1) \leq f(k_2)$), and since $\{k\}$ is finite, it follows that $k = f(k)$ for each k.

If $|\{k\}| = 1$, then, by the above, all lines and points of π_0 have the same degree k, say. So π_0 is a possibly degenerate projective plane of order $k - 1$, and S is $PG(3,k - 1)$ or a set of four points with $k = 2$.

Suppose $|\{k\}| \geq 2$, and let L be a line of maximal size $k \geq 3$ in π_0. All points of $S - L$ have degree $\geq k$ and so, by the above, have degree k. If $\pi_0 - L$ contains at least two points, p, q, then at least two points of π_0 have degree k and since the sets of line degrees and point degrees of π_0 are equal it follows that π_0 contains two k-lines which must necessarily intersect. Let x be the point of intersection. Then x is the only point of π_0 which can have degree less than k. Suppose it has degree k'. Then the set of line sizes is $\{k, k'\}$ where, by assumption $k' < k$. From $k = f(k)$ above we also have that every point of S is on precisely one line with k' points, thus k'-lines on π_0 do not intersect. Let $y \neq x$ in π_0. Then y is on a k-line M which is not on x, since $k \geq 3$ implies y is on at least two k-lines. Now in $\pi_0 = \langle x,M \rangle$, x has degree at least k, which is a contradiction. We conclude that π_0 contains only a single point outside L and is therefore a degenerate projective plane with $k + 1$ points. Using the second claim above, each point of S is then on a unique k-line and k 2-lines. It follows that S is the disjoint union of two k-lines. □

At this point, we wish to mention a result which does not fall within the scope of this text for the simple reason that it is a result about infinite systems. Nevertheless, since it furnishes an additional example of 'π-space' beyond what we have seen so far, it is, in our view, important enough to mention. We do not give the proof.

Theorem 7.3.4. (Ceccherini and Tallini 1986.) Let S be a d-dimensional projective (affine) space constructed over an infinite, countable field K, $d \geq 3$. Let $\bar{d}$ be an integer such that $2 \leq \bar{d} \leq \dfrac{d+1}{2}$. Then there exists a set of $\bar{d}$-spaces in S such that each plane is in a unique $\bar{d}$-space. Moreover, the space S equipped with this set of $\bar{d}$-spaces as planes is a π-space in which each plane is isomorphic to the same $\bar{d}$-space of the projective geometry.

The above examples of π-spaces are essentially all designs. It is not known whether or not 'interesting' π-spaces exist when π has at least two line sizes. However, the next result tells us that there cannot be many such spaces. We show that if each plane of the dimensional finite linear space S is isomorphic to the same plane π in which line sizes have at least two values, then there can be only a finite number of such spaces S. The examples of affine and projective spaces indicate that this result does not hold for the case of a single line size. The theorem is due to F. Buekenhout and R. Deherder (1971).

Theorem 7.3.5. Let π be a linear space with at least two lines of different sizes. Then there exists a finite number only, of finite dimensional linear spaces in which each plane is isomorphic to π.

PROOF. We begin by introducing a little notation which will be more convenient for this proof. We retain v as the number of points in S, but in addition, let w be the number of points in π. Let also

I be the set of line sizes of S (and of π),
b_i be the number of lines of S with i points,
β_i be the number of lines of π with i points, and
c be the number of planes of S.

Proposition 1.5.3 then gives

$$\sum_{i \in I} b_i i(i-1) = v(v-1).$$

Counting in two different ways the pairs consisting of a line of i points and the planes on that line, we obtain also

$$c\beta_i = b_i \frac{v-i}{w-i}.$$

Combining these two equations yields

$$c \sum_{i \in I} \beta_i i(i-1) \frac{w-i}{v-i} = v(v-1).$$

We then introduce three functions $P(v)$, $P_i(v)$ and $Q(v)$ as follows:

$$P(v) = \sum_{i \in I} (v-i), \qquad P_i(v) = \frac{P(v)}{v-i},$$

$$v(v-1)P(v) = c \sum_{i \in I} \beta_i i(i-1)(w-i)P_i(v) = cQ(v).$$

Hence $Q(v) \mid v(v-1)P(v)$.

As polynomials in x, $Q(x)$ and $x(x-1)P(x)$ have degrees respectively $|I|-1$ and $|I|+2$. We consider the roots of $x(x-1)P(x)$. Clearly 0 and 1 are roots, as is any element of I. But each element of I is at least 2, and there are at least two elements in I. No element of I is a root of $Q(x)$, and neither is 0 or 1. So we have four distinct roots of $x(x-1)P(x)$ which are not roots of $Q(x)$. Since the degrees of the functions differ only by 3, it follows that $Q(x)$ does not divide $x(x-1)P(x)$. We can therefore write

$$x(x-1)P(x) = Q(x)D(x) + R(x)$$

where $R(x)$ is a non-zero polynomial with degree $< |I| - 1$, and $D(x)$ and $R(x)$ have rational coefficients. Thus, there is some integer λ such that

$$\lambda x(x-1)P(x) = \lambda Q(x)D(x) + \lambda R(x)$$

is an equation with integer coefficients. Since $Q(v)|v(v-1)P(v)$, it follows that $Q(v)|\lambda R(v)$. But as v gets larger, the quotient $\lambda R(v)/Q(v)$ tends to zero, since the degree of $R(x)$ is less than the degree of $Q(x)$. Hence there can be only a finite number of values of v for which the above quotient is an integer. □

More information on π-spaces can be found in Delandtsheer (1982) or in Leonard (1982).

7.4 Local structures

We now formalize a concept that we have in fact already encountered more than once before this section. This is the idea of a linear space having 'locally' a certain type of structure.

Let S be a linear space. For each point p of S, we let S_p denote the structure of *all* subspaces of S which contain p. In general, S_p need not be a linear space. (See the exercise section.) However, it is easy to see that if S is a dimensional linear space, then so is S_p for each point p of S.

The class of linear spaces we wish to study primarily in this section is that in which S_p is a projective space for each point p. We call these *locally projective (linear) spaces*. The following result is not difficult to prove in this case, and we leave the demonstration as an exercise.

Proposition 7.4.1. Let S be a linear space in which for each point p, the structure S_p is that of a projective space. Then S is a dimensional linear space; moreover, S_p is isomorphic to S_q, for each q, which implies in particular that the orders of the projective spaces at each point are the same.

The first major result of this section is an elegant result of J. Doyen and X. Hubaut (1971), who completely classify locally projective finite linear spaces of dimension greater than 3 having the property that all lines have the same number of points.

Theorem 7.4.2. A finite locally projective space of dimension greater than 3 in which all lines have n points is either a projective space or an affine space.

PROOF. Using Proposition 7.4.1, we may let k be the common order of each projective space S_p, and v_2 and v_3 respectively the number of points in a plane and in a 3-space of S.

Clearly S is a 2-$(v,n,1)$-design, and so Proposition 1.6.1 gives

$$\frac{v-1}{n-1} = \frac{k^d - 1}{k - 1}.$$

The fact that S is locally projective yields the equations

$$v_2 = (k+1)(n-1) + 1 \qquad \text{and} \qquad v_3 = (k^2 + k + 1)(n-1) + 1.$$

Let $m = k + 1 - n$. Then S is projective precisely when $m = 0$. Suppose then that $m \geq 1$. It follows that for any plane π, we can find a line L missing π, but in a 3-space with π. Let p be a point of π. Then $\langle p,L\rangle \cap \pi$ is a line, since S is locally projective. But this induces a spread in π, and we may conclude that $n|v_2$. It follows from the equation above for v_2, that $n|k$. A similar argument results in $v_2|v_3$.

Let $k = xn$, x an integer ≥ 1. We can write $v_3 = n(xv_2 - x + 1)$. It follows that $v_2|(x-1)n$; but since $v_2 = n(xn - x + 1)$, this gives $xn - x + 1|x - 1$. Now we can write $n = (xn - x + 1) - (x - 1)(n - 1)$. Hence $xn - x + 1|n$, implying $x(n-1) + 1 \leq n$ and so $x(n-1) \leq n - 1$. It follows that $x = 1$, since $n - 1 \geq 1$.

Since $x = 1$ we have $m = 1$, and it is now easy to show that the axioms for an affine space (Section 1.2) are satisfied. □

The reader is invited to locate the sole place in the above proof at which $d > 3$ is used.

Theorem 7.4.3. A finite locally projective space of dimension 3 in which all lines have n points is a projective space, an affine space or a space in which each plane is I-affine, where $I = \{n^2 - n + 1\}$ *or* $I = \{n^3 + 1\}$.

PROOF. Through each point there are $k^2 + k + 1$ planes, and so the total number of planes is $v(k^2 + k + 1)/v_2$, an integer. But $v = k^2(n - 1) + v_2$, which implies $v_2 | k^2(k^2 + k + 1)(n - 1)$. Since $v_2 = (k + 1)(n - 1) + 1$ we see that $(v_2, n - 1) = 1$ and $v_2 | k^2(k^2 + k + 1)$. Then from $k = (v_2 - n)/(n - 1)$, we derive $v_2 | n^2(n^2 - n + 1)$.

We now take a result from the proof of Theorem 7.4.2 which used only dimension ≥ 3. That is, if S is not projective, then $n | v_2$ and $n | k$. Also, putting $k = xn$, we then obtain $v_2 = n(xn - x + 1)$, and so $xn - x + 1 | n(n^2 - n + 1)$. Thus there exists a non-negative integer y such that

$$\begin{aligned}(x(n - 1) + 1)(y(n - 1) + 1) &= (n^2 + 1)(n - 1) + 1\\ &= n(n(n - 1) + 1).\end{aligned}$$

If $y = 0$, we obtain $x = n^2 + 1$. Otherwise, $1 \leq x < n^2 + 1$, and so $1 \leq y \leq n$; but then $x \leq n$. Rewriting the above equation as

$$xy(n - 1) + x + y = (n + 1)(n - 1) + 2$$

results in

$$x + y \equiv 2 \pmod{(n - 1)}.$$

The only possible cases now are (a) $x + y = 2$, $xy = n + 1$; (b) $x + y = n + 1$, $xy = n$; (c) $x + y = 2n$, $xy = n - 1$. The cases (a) and (c) are easily eliminated. The solutions of (b) are $x = n$, $y = 1$ and $x = 1$, $y = n$.

The three values $x = 1$, n and $n^2 + 1$ obtained above result in $k = xn = n$, n^2 or $n^3 + n$. Thus S is I-affine with $I = \{1\}$, $\{n^2 - n + 1\}$ or $\{n^3 + 1\}$. □

We leave the difficult question of examples of spaces of the above types as a research problem.

The restriction above, that all lines have the same size, is quite a severe one, and it is what allows us to do the very neat counting and divisibility arguments. Can we say anything if this restriction is dropped? Certainly the type of argument would have to be different, as we can no longer rely simply on counting. At the most optimistic, the above results might indicate that for dimension > 3, a finite locally projective space embeds in projective space. In fact, this *is* the case, as we shall show below. For dimension 3, however, we run into some problems. In the case of an $\{n^2 - n + 1\}$-affine linear space with $n = 2$, we have a 3-dimensional linear space in which each line has 2 points, each plane 6 points and the whole space 22 points. It can be shown that this linear space does *not* embed in a 3-dimensional projective space of order 4. But under additional conditions in the dimension 3 case, we can get an embedding theorem.

The important results of the last twenty years or so on the embedding question have considered more general structures than linear spaces, while a number of earlier papers did indeed consider the embedding problem for linear spaces. In 1953 O. Wyler had already considered locally projective linear spaces of dimension ≥ 3 satisfying assumptions equivalent to our two major conditions I and II below. We refer the reader to his paper, listed in the bibliography, and also to the references in that paper. Since the later papers deal with much more general settings than ours, we give only some references: L. M. Batten (1978a,b, 1987a), R. Frank (1985), A. Herzer (1976, 1979, 1981), J. Kahn (1980a,b), W. M. Kantor (1974), R. Wille (1967, 1971).

The proof presented below is based primarily on Batten (1987a) but is essentially also that of Batten (1978b), Kantor (1974) and Wille (1967) when particularized to the linear space cases. Moreover, we allow in this theorem the slightly more general situation of a *generalized* projective space at each point. Note that the finiteness of S plays no role in the proof, and hence can be dropped.

Theorem 7.4.4. Let S be a locally generalized projective linear space of dimension > 3. Then there is an embedding ϕ of S into a generalized projective space such that for every subspace V of S, V and $\phi(V)$ have the same dimension.

PROOF. We show first that in such a space the following holds:

I (Bundle Theorem) Let L_1, L_2, L_3, L_4 be lines, no two intersecting, no three coplanar, such that five of the six pairs of lines are coplanar. Then the sixth pair is coplanar.

Let the lines L_1, L_2, L_3, L_4 be as above where all pairs are coplanar except possibly L_3 and L_4. Then $V = \langle L_1,L_2,L_3\rangle$ and $W = \langle L_1,L_2,L_4\rangle$ have dimension 3. If L_3 and L_4 are not coplanar, then $\langle L_3,L_4\rangle = \langle L_1,L_2,L_3,L_4\rangle$ has dimension 3. This forces $V = W$, implying $L_4 \subseteq V$. Let $p \notin V$. Define $L_p = \langle L_1,p\rangle \cap \langle L_2,p\rangle$, and $L_p \not\subseteq V$. By the above argument, $\langle L_3,L_p\rangle$ and $\langle L_4,L_p\rangle$ are planes, and so $\langle L_3,L_4,L_p\rangle$ has dimension 3. Since $\langle L_3,L_4\rangle$ has dimension 3, this forces the contradiction $p \in V$. This completes the proof of condition I.

For parallel lines L and L' and any point $p \notin \langle L,L'\rangle$, define $L_p =$

$\langle L,p\rangle \cap \langle L',p\rangle$. Fix a point $q \notin \langle L,L'\rangle$. For $p \in \langle L,L'\rangle$, $p \notin L$, define $L_p = \langle L,p\rangle \cap \langle L_q,p\rangle$. For $p \in L$, define $L_p = L$. By condition I above, L_p for $p \in \langle L,L'\rangle$ is seen to be independent of the choices of q. Thus the set

$$[L,L'] = \{L_p \text{ as defined above}\}$$

is a parallel class of S.

If $x = [L,L']$ and $p \in S$, by px (or xp) we mean the unique line of x on p.

We now wish to prove the following condition:

II Let p_1, p_2 be points, π_1, π_2 planes and L_{ij}, $1 \le i \le 2$, $1 \le j \le 3$ be lines, such that for $1 \le i \le 2$, $p_i \in L_{ij} \subseteq \pi_i$, $1 \le j \le 3$. Suppose $\pi_1 \cap \pi_2 = \phi$ and that $L_{1j} \| L_{2j}$ for $1 \le j \le 3$. Then for $p \notin \pi_1$ or π_2, setting $x_j = [L_{1j},L_{2j}]$, the lines px_j are all coplanar, $1 \le j \le 3$.

Suppose p_i, L_{ij} and π_i are as above, and suppose $p_3 \notin \langle \pi_1,\pi_2\rangle$, which has dimension 3. Then $\langle \pi_i,p_3\rangle$ has dimension 3 for $1 \le i \le 2$. Let L_{3j} be the line p_3x_j. If $\langle L_{31},L_{32},L_{33}\rangle$ has dimension 3, we obtain $\langle L_{31},L_{32},L_{33}\rangle \subseteq \langle \pi_i,p_3\rangle = \langle \pi_1 \cup \pi_2\rangle$ for $i = 1, 2$, and so $p_3 \in \langle \pi_1,\pi_2\rangle$, a contradiction. If $p_3 \notin \pi_1$ or π_2, but $p_3 \in \langle \pi_1,\pi_2\rangle$, choose $p_4 \notin \langle \pi_1,\pi_2\rangle$, and define $L_{4j} = \langle L_{1j},p_4\rangle \cap \langle L_{2j},p_4\rangle$, $1 \le j \le 3$. Then $p_3 \notin \langle \pi_1,p_4\rangle$. By the above argument, the lines L_{4j}, $1 \le j \le 3$, are coplanar, and by condition I, $L_{3j} = \langle L_{1j},p_3\rangle \cap \langle L_{4j},p_3\rangle$, $1 \le j \le 3$. Now once again using the above argument, the L_{3j}, $1 \le j \le 3$, are coplanar. This completes the proof of condition II.

We now introduce a new structure S^* in which the points are all the points of S along with all sets $[L,L']$. For any point x of S^* we write xIM, M a line of S, if $x \in M$ or if $M \in x$. We say *x is incident with M*. For any subspace V of S, we write xIV precisely when x is incident with some line of V.

For distinct planes π and π' of S such that $\langle \pi,\pi'\rangle$ has dimension 3, define a *line* of S^* to be

$$[\pi,\pi'] = \{x | xI\pi \text{ and } xI\pi'\}.$$

We shall refer to such a line as $\pi \cap \pi'$.

We prove now that the points and lines of the new structure S^* are the points and lines of a generalized projective geometry.

We note first that if p is a point of S in a plane π and x a new point incident with π, then the line M of S corresponding to px in S^* is a line of π. To see this, let L be a line of π with which x is incident, and let

q be an old point of this line. If p is on L we are finished. Otherwise, M and L are distinct and coplanar, and so $\langle M,L\rangle = \pi$. Thus $M \subseteq \pi$.

It is not difficult to see that each line of L^* has at least two points, and that two points of S^* are on at least one common line.

To prove that any two points of S^* are on a unique line, it suffices to prove that if x, $y\mathrm{I}\pi$, π' and π'', and if $z\mathrm{I}\pi$ and π', then $z\mathrm{I}\pi''$.

Suppose $\pi \cap \pi' = L$, a line of S. If there is a point q of L and of S in π'', then at least two of qx, qy and qz are lines of π, π' and π''. So they cannot be distinct. It follows that $LI\pi''$, and so $zIL \subseteq \pi''$ implies $zI\pi''$ as desired. If $L \cap \pi'' = \phi$ in S, let $p \in \pi'' - L$, $p \in S$. Then px and py are lines of π'' and each is parallel to L. If $px \neq py$, then $\langle L,p\rangle = \langle px,py\rangle = \pi''$, implying the contradiction $L \subseteq \pi''$. So $px = py$. It follows that $x = [L,px] = [L,py] = y$, a contradiction.

Suppose then that $\pi \cap \pi' = \phi$ in S. Then there are points p, p', p'' of π, π', π'' respectively. Applying condition II to the lines px, py, pz, $p'x$, $p'y$, $p'z$, $p''x$, $p''y$, $p''z$, we obtain the result that the line of S corresponding to $p''z$ lies in π''. This implies $zI\pi''$, as desired.

For any distinct points p and q of S^* we now let pq denote the unique line of S^* on p and q.

In order to conclude the proof of the theorem, we use an alternate definition of generalized projective geometry. It is a linear space in which the Veblen–Young axiom holds: If p, q, x and y are points such that the line on p and q and the line on x and y have a point in common, then the line on p and x and the line on q and y have a point in common.

Suppose first of all that L and L' are lines lying in a plane π of S. If both are lines of S, then by construction, they now meet in S^*. If L, say, is new, it corresponds to at least two distinct planes of S and therefore we can find a point x of S not in π. Then $\langle L,x\rangle \cap \langle L',x\rangle = M$ is a line of S. Let $p \in \pi$, $p \in S$, and define $L'' = \langle p,M\rangle \cap \pi$. Then L'' is a line of S. So $M \cap L'' = z$ is a point of S^*. But $zI\pi \cap \langle L,M\rangle = L$ also, and similarly zIL'.

Now let p, q, r and s be points as in the Veblen–Young axiom such that $tIpr$ and $tIqs$. We wish to show that the line pq meets the line rs in S^*. We may assume that these five points do not lie in a common plane of S. There is some point x of S such that $\langle x,p,q\rangle$ and $\langle x,r,s\rangle$ are distinct planes of S. It follows that $\langle x,p,q\rangle \cap \langle x,r,s\rangle = L$. Then pq and L lie in a common plane of S and thus, by the first stage of this proof, meet in a point z of S^*. Similarly, $rs \cap L = z'$ in S^*. If $z = z'$ we are finished. So suppose $z \neq z'$.

For a set of points $x_1, x_2, \ldots, x_m$ of S^*, define $(x_1, x_2, \ldots, x_m)$ to be the intersection of all sets of points X of S containing $x_1, x_2, \ldots, x_m$ with the property that if x and y are points of X then each point of the line xy is in X. Thus $(p,q,r,s,t) = (t,z,z') = (t,L) \subseteq \langle t,L \rangle$. But $\langle t,L \rangle$ is a plane of S, which contradicts our assumption on the points p, q, r, s and t.

It is now straightforward to check that the above construction satisfies the embedding conditions of the theorem. □

The above proof in fact does not use all the axioms of a linear space, and we leave it to the reader to find the most general system for which it still goes through.

As already mentioned the restriction of dimension > 3 is essential in the proof. However, J. Kahn's result (1980a) implies that under the additional assumption that the Bundle Theorem (BT) holds, the embedding works for dimension 3. We state this particular instance of his theorem as a theorem here, for the sake of completeness. Note, however, that he needs the space to be *locally projective*.

Theorem 7.4.5. Let S be a locally projective 3-dimensional linear space satisfying BT. Then S is embeddable in a unique way in PG (3,n) where n is the order of the projective plane at some (and therefore at each) point of S.

Interestingly enough, A. Herzer (1979, 1981) proves the same result if condition II of the above proof holds, rather than condition I. He needs additionally, however, that the n of Theorem 7.4.5 be at least 4.

We make one final comment before leaving locally projective spaces. For linear spaces, allowing generalized projective spaces at each point is in fact not much more general than allowing only projective spaces at each point. L. M. Batten (1987b) shows that if S is a locally generalized projective linear space of dimension ≥ 3 then it is either locally projective or is itself generalized projective.

A natural question to ask now is: What can be said of a linear space S if for every point p, the structure S_p is an affine space? In fact if S has finite dimension $d \geq 4$, it follows from one of the main results of W. M. Kantor's paper (1974) that S embeds in a d-dimensional generalized projective space. For dimension 3, the results in the literature have generally appeared in the guise of theorems about *inversive spaces*, which we shall not define here, but which can be considered to be linear spaces in which all lines have precisely two points. The most significant information and results on these geometries can be found in P. Dembowski (1968) and J. Kahn (1980b; 1982).

7.5 'Global' conditions

All results currently existing in the literature have essentially to do with line sizes. Thus we give only one result in this section. Compare the statement of Theorem 7.5.1 below with that of Theorem 3.3.6.

A *plane k-arc* of a d-dimensional linear space is a k-arc of some plane of the space.

It is well known (see Dembowski 1968, for instance) that an $(n + 1)$-arc in $PG(2,n)$ can be completed to an $(n + 2)$-arc precisely when n is even. The additional point is obtained by adding the *knot* of the $(n + 1)$-arc, that is, the unique point which is on all the tangents to this arc.

A *quadric* in $PG(d,n)$, $d \geq 0$, is the set of points satisfying a second degree homogeneous polynomial equation in $d + 1$ variables over $GF(n)$. The number of points of a non-degenerate quadric Q is $(n^{d-1} + \ldots + 1) - n^{(d-1)/2}$, $n^{d-1} + \ldots + 1$ or $(n^{d-1} + \ldots + 1) + n^{(d-1)/2}$ depending on whether Q is *elliptic*, *parabolic* or *hyperbolic* respectively. (See L. M. Batten 1986 or J. W. P. Hirschfeld 1979 for more details.) The *nucleus space* of a quadric Q in $PG(d,n)$ is the set of points of $PG(d,n)$ through each of which there is no line meeting Q in just two points.

Let R and R' be disjoint subspaces of S, and K and K' respectively subsets of the point sets of R and R'. Then the *cone determined by* K and K' is the set of all points on the lines pp', $p \in K$, and $p' \in K'$.

Theorem 7.5.1. Let S be a d-dimensional linear space, $d \geq 3$, and $n \geq 2$ an integer such that

(1) $k_L \in \{n - 1, n, n + 1\}$ for all lines L,
(2) $r_p^i = n^{i-1} + \ldots + 1$ for all points p in any i-space.

Then S is embeddable in a unique way in $PG(d, n)$. If, in addition,

(3) $v \leq n^d$, then S is the complement in $PG(d,n)$ of one of the following:

(i) A (possibly singular) parabolic quadric
(ii) A (possibly singular) hyperbolic quadric
(iii) A parabolic quadric plus a subspace of its nucleus space
(iv) A cone projecting from a $PG(d - 3, n)$ a plane $(n + 1)$-arc plus a subspace of $PG(d - 2, n)$ joining the knot of the arc with the $PG(d - 3, n)$

(v) A hyperplane and a subspace of $PG(d,n)$.

In order to prove the above we make use of a specialized theorem which we state here without proving.

Theorem 7.5.2. (Tallini 1956.) Let X be a set of points in $PG(d,n)$, $d \geq 3$, $n > 2$, such that $|X \cap L| \in \{0,1,2,n+1\}$ for all lines L, and $|X| \geq n^{d-1} \ldots + 1$. Then X is one of the sets (i)–(v) of Theorem 7.4.1.

Note that if X is as in Tallini's theorem, then $S - X$ satisfies (1), (2) and (3) of Theorem 7.5.1.

We break the proof of Theorem 7.5.1 into several propositions.

Proposition 7.5.3. Let S be a d-dimensional linear space, $d \geq 3$, satisfying condition (2) of Theorem 7.5.1. Then

(i) an i-space is uniquely determined by any $i + 1$ of its points not all in an $(i - 1)$-space;

(ii) each line is on $n^{i-2} + \ldots + 1$ planes in any i-space;

(iii) any two distinct planes in a 3-space meet in ϕ or a line.

PROOF. Part (i) follows from the definition of d-dimensional linear space. For part (ii), let L be a line of an i-space V and p a point of L. Each plane on L has $n + 1$ lines on p by condition (2). Since there are $n^{i-1} + \ldots + 1$ lines on p in V by condition (2), there must be precisely $n^{i-2} + \ldots + 1$ planes on L in V. Suppose finally, that in some 3-space the planes π and π' meet in a single point p. Let L be a line on p in π. Using (2), the $n + 1$ lines on p in π' form $n + 1$ planes on L because of part (i). But π is an additional plane on L, contradicting part (ii). □

Part (iii) of proposition 7.5.3 now implies that S is locally projective. In case $d > 3$, we invoke the results of the previous section to get an embedding in a unique way in $PG(d,n)$. If $(n - 1)$-lines exist, then clearly $n > 2$ and Tallini's theorem gives the desired result. If $(n - 1)$-lines do not exist (and $d \geq 3$), it is easy to see that each plane of S is affine, projective or projective less a point. If all planes are affine, then S is affine (we are *not* using Theorem 7.3.1 here!) and we have (v) of Tallini's theorem. Fix an $(n + 1)$-line. By part (ii) of Proposition 7.5.3, this is on $n^{d-2} + \ldots + 1$ planes, each plane having either $n^2 + n + 1$ or $n^2 + n$ points. It follows that $v \geq (n^2 - 1)(n^{d-2} + \ldots + 1) + n + 1 = (n + 1)n^{d-1} = n^d + n^{d-1}$, contradicting condition (2).

It suffices therefore to restrict ourselves to the case $d = 3$. Moreover, by the above, we may assume that $(n - 1)$-lines exist, and hence $n > 2$.

Proposition 7.5.4. Let S be a d-dimensional linear space, $d \geq 3$, satisfying conditions (1) and (2) of Theorem 7.5.1. Then no plane of S contains just $(n - 1)$-lines.

PROOF. Suppose the plane π is only on $(n - 1)$-lines. By condition (2), each point of π is on $n + 1$ lines in π. Hence $b_\pi = [(n + 1)(n - 2) + 1](n + 1)/(n - 1)$, implying $n - 1|2$, and so $n = 3$. In this case, it is easy to see that some line L misses π and so induces a parallel class of π into 2-lines using part (iii) of Proposition 7.5.3. This contradicts $v_\pi = 5$. □

Proposition 7.5.5. Let S be as in Proposition 7.5.4. If an $(n - 1)$-line M is parallel to some n-line, then M determines precisely two parallel classes of S. In particular, if a parallel class is not on an n-line in a plane π, then it is on $n + 1$ $(n - 1)$-lines of π. It follows that BT of Theorem 7.4.5 is satisfied for any four lines of a parallel class on M.

PROOF. Clearly, condition (2) implies that M is on at most two parallel classes. Moreover, it follows from (2), and (i) of Proposition 7.5.3 that any n-line determines a unique parallel class of S. Let L be an n-line parallel to M, and let $[M]_1$ be the parallel class on M determined by L. Consider the plane $\pi = \langle M, L\rangle$. If some point of L is on a second n-line of π missing M, we are finished. Suppose then, that each point of L is on an $(n - 1)$-line M_i missing M. If $\{p\} = M_i \cap M_j$, $i \neq j$, there is a unique line of $[M]_1$ on p missing M and L, and we contradict (2); so the M_i's are pairwise parallel. Let $q \in \pi - \langle L \cup M\rangle$. By (2), there is a line on q missing M but meeting L. Thus this line must be on some M_i. It follows that the M_i along with M form a parallel class of π.

Let p be any point outside π and let L' be the unique element of $[M]_1$ on p. Let M' be the second line of $\pi' = \langle M,L'\rangle$ on p missing M. Using (iii) of Proposition 7.5.3, it is clear that $\langle M_i,p\rangle \cap \pi' = M'$ for all i. It follows that a parallel class $[M]_2$ of S is induced on M. □

Proposition 7.5.6. If the conditions of Theorem 7.5.1 hold with $d = 3$, $n \geq 2$, and if there are no $(n + 1)$-lines, then S is a projective geometry PG(3,n) less a plane, less a plane and a line not in the plane, or less two planes.

PROOF. Let π be a non-affine plane on a fixed point p. Letting a and c be respectively the number of $(n-1)$- and of n-lines on p in π, we obtain $v_\pi = a(n-2) + c(n-1) + 1$. On the other hand, π contains at least one n-line, by Proposition 7.5.4, which thus determines a parallel class of π. Letting d and e respectively be the number of $(n-1)$- and of n-lines in this parallel class, we find that $a(n-2) + c(n-1) + 1 = d(n-1) + en$. Consequently, $n - 1 | a + e - 1$, while $1 \leq a \leq n$ and $1 \leq e \leq n + 1$. So $1 \leq a + e - 1 \leq 2n$. It follows that $a + e - 1 = n - 1$ or $2(n-1)$, or, in case $q = 3$, $3(n-1)$. Then $a + c = d + e + i$ where $i = 1$ or 2, or if $n = 3$, $i = 3$. Since $a + c = n + 1$, this implies $d + e = n$, $n - 1$ or $n - 2$. Clearly this last is not possible. Hence any parallel class which is determined by an n-line has either n or $n - 1$ elements in any plane on that n-line.

Suppose the plane π contains an n-line L disjoint from an $(n-1)$-line M. From the above, the parallel class in π on M and not on L contains $n + 1$ $(n-1)$-lines. Thus $v_\pi = n^2 - 1$. Fixing a point and counting $(n-1)$- and n-lines on it, we see that each point is on a unique $(n-1)$-line. We thus get a single partition of π into $n + 1$ $(n-1)$-lines, and $n + 1$ partitions of π into $n - 1$ n-lines and one $(n-1)$-line.

Now suppose that in π all $(n-1)$-lines meet all n-lines. Any n-line induces a parallel class of π which, by the argument above, has at most n elements. Since the parallel class induced contains only n-lines, it follows that either π is an affine plane, or it contains a unique parallel class of $n - 1$ n-lines, all other lines being $(n-1)$-lines, and $v_\pi = n^2 - n$.

So there are three possible types of plane π in S which we shall refer to as types I, II and III, with $v_\pi = n^2$, $v_\pi = n^2 - 1$ and $v_\pi = n^2 - n$ respectively.

Fix an $(n-1)$-line M. Let s be the number of planes on M of type III. Since there can be no planes on M of type I, we obtain $v = s(n^2 - 2n + 1) + (n + 1 - s)(n^2 - n) + (n - 1)$ using part (ii) of Proposition 7.5.3.

Suppose M is on no plane of type II; that is $s = n + 1$. Then $v = n^3 - n^2$. Thus *each* $(n-1)$-line is only on planes of type III. Fix an n-line L now. Clearly, each plane on L is either of type I or of type III. Counting as for M above, we find that L is on a unique affine plane, and on n planes of type III. It follows that each affine plane induces a parallel class of affine planes in S in which there must be $(n^3 - n^2)/n^2 = n - 1$ planes. Hence, there is a unique such parallel class, and all lines not in one of these planes are $(n-1)$-lines, and each $(n-1)$-line meets each affine plane.

Let M_1 and M_2 be parallel $(n-1)$-lines. Let p be a point not in $\langle M_1,M_2\rangle$; let $M_p = \langle M_1,p\rangle \cap \langle M_2,p\rangle$, a line by (iii) of Proposition 7.5.3; and let A be the unique affine plane on p. It is not difficult to see that M_p is parallel to each element of the parallel class induced in $\langle M_1,M_2\rangle$ (a plane of type III) by M_1 and M_2.

Let M_q be a fourth line, parallel to M_1 and M_2 and not in $\langle M_1,M_2\rangle$, $\langle M_1,M_p\rangle$ or $\langle M_2,M_p\rangle$. Let $q = M_q \cap A$, $u = M_1 \cap A$, $v = M_2 \cap A$. Thus the lines uv and pq are distinct in A. If $uv \cap pq = \{z\}$, let M be on the line of z of the parallel class on M_1 and M_2 in $\langle M_1,M_2\rangle$. Then both M_p and M_q are coplanar with M, and so $M_p, M_q \subset \langle M,p,q\rangle$ implying $M_p \| M_q$. If $uv \| pq$ in A, take a second affine plane $A' \neq A$, and define correspondingly in the obvious manner the points p', q', u', v'. If $p'q'$ meets $u'v'$, the above argument using A' yields $M_p \| M_q$. So suppose $p'q' \| u'v'$. But an n-line uniquely determines a parallel class of n-lines in S, and so it follows that $pq \| p'q'$ forcing $M_p \| M_q$.

Hence if $s = n + 1$, BT of Kahn's theorem holds in S. If $s < n + 1$, each $(n-1)$-line is on a plane of type II and so is parallel to some n-line. By Proposition 7.5.6, it again follows that BT holds in S. By Kahn's theorem then, S is embeddable in $PG(3,n)$. Then by Tallini's theorem, S is one of (i)-(v).

In case $s = n + 1$ above, and $v = n^3 - n^2$ with a unique parallel class of $n - 1$ affine planes, it is now easy to see that S is the complement of two planes in $PG(3,n)$.

In case $s < n + 1$ and planes of type II exist, we see that S is the complement of a plane and a line not in the plane in $PG(3,n)$. □

The remainder of the proof of Theorem 7.5.1 now follows from the next proposition.

Proposition 7.5.7. If the conditions of Theorem 7.5.1 hold with $d = 3$, $n \geq 2$, and if $(n + 1)$-lines exist, then S is a projective geometry $PG(3,n)$ less a hyperbolic quadric consisting of two partial parallel classes of $n + 1$ lines each such that each line of one meets all lines of the other (a grid), or less a set of $n + 1$ or $n + 2$ lines all on a common point (a cone). (In the last case, q must be even.)

PROOF. Suppose first of all that there exists an $(n-1)$-line M not parallel to any n-line. We determine the types of planes that can exist on an $(n-1)$-line.

Let π be a plane on the $(n-1)$-line M. By Proposition 7.5.4, π contains an n- or $(n + 1)$-line. Since any line parallel to M but not in π induces a parallel class of π on M, such a parallel class has either n or $n + 1$ $(n - 1)$-lines. So $v_\pi = (n - 1)n$ or $(n - 1)(n + 1) = n^2 - 1$. In the first case, there is no $(n + 1)$-line in π, there is a unique parallel class of n-lines, and π is the complement of two lines in a projective plane of order n. We say π is of type I. In the second case, there is no n-line (this would induce a parallel class of n-lines), each point is on $(n + 2)/2$ $(n - 1)$-lines, and π is the complement of an $(n + 2)$-arc in a projective plane of order n. We say π is of type II.

We prove now that if M is an $(n - 1)$-line not parallel to any n-line, then *no* $(n - 1)$-line is parallel to any n-line.

Let π be any plane such that $M \cap \pi = \phi$. Then M induces a parallel class of lines in π. For each line M' of this parallel class, the plane $\langle M,M' \rangle$ is of type I or II, and so M' is an $(n - 1)$-line. Since by Proposition 7.5.4, π contains an n- or $(n + 1)$-line, it follows that π is of type I or II, or possibly a third type of plane with $v = n^2 - 1$ in which there is no $(n + 1)$-line and in which there is a unique parallel class of $n + 1$ $(n - 1)$-lines. π is the complement of a line and a point not on the line in a projective plane of order n. In the plane $\langle M,M' \rangle$ there is a second $(n - 1)$-line meeting M and parallel to M'. This line also generates a parallel class of $(n - 1)$-lines in π containing M'. Thus π cannot be the new type of plane above, but must be of type I or II.

Now let L be an arbitrary $(n - 1)$-line distinct from M. If $L \| M$, then any plane on L is of type I or II using the above argument.

If L meets M, then $\langle L,M \rangle$ is of type I or II and so contains an $(n - 1)$-line M' parallel to both L and M. By the above, M' is only on planes of type I or II, and transitivity now implies that this is also true for L.

If $\langle L,M \rangle$ is S, then not all planes on L meet M. Choose one, π, which does not. Let $p \in \pi - L$. Then $\pi \cap \langle M,p \rangle = M'$ is an $(n - 1)$-line parallel to M. Thus M' is only on planes of type I or II. Now either M' is parallel to L or M' meets L. By the above, L is only on planes of type I or II.

It follows that no $(n - 1)$-line is parallel to any n-line.

We now determine the possible planes on an n-line.

Consider an arbitrary n-line L in a plane π. Since L induces a parallel class of $n - 1$, n or $n + 1$ lines, and since no $(n - 1)$-line is parallel to L, π is of type I, or an affine plane, or a projective plane less a single point. However, an affine plane clearly determines a parallel class of at least $n + 1$ affine planes (since $(n + 1)$-lines exist). If there is an $(n - 1)$-line M, it therefore misses some affine plane A. Let $p \in A$. Then $\langle M,p \rangle \cap A = L$ is an n-line by (iii) of Proposition 7.5.3, and L and M are parallel, which contradicts the above. So $(n - 1)$-lines do not exist in this case. We may now count v using the planes on an n-line and then on an $(n + 1)$-line as follows, letting a be the number of affine planes on an n-line: $v = a(n^2 - n) - (n + 1 - a)n^2 + n = n^3 + n^2 - na$, and $v = (n + 1)(n^2 - 1) + (n + 1) = n^3 + n^2$. It follows that $a = 1$, and each n-line is on a unique affine plane. Counting v again, using the number of lines on a fixed point, we see that each point is on $n + 1$ n-lines. Thus each n-line belongs to a plane of the parallel class of precisely $n + 1$ affine planes. Adding new points in the obvious manner, we see that S is a projective 3-space less a single line with all its points. We shall say that the 'punctured' projective plane is of type III. We may now assume that affine planes do not exist.

Finally, consider an arbitrary $(n + 1)$-line L in a plane π. If π also contains an $(n - 1)$-line, then it must be of type II; if it contains an n-line, it must be of type III. Suppose it is projective. Let M be any $(n - 1)$-line. Then it meets π, and all planes on M are necessarily of type II. It follows that $v = n^3 - 1$. Since any n-line generates a parallel class of n-lines, $n|v$ giving a contradiction. So n-lines do not exist and hence nor do planes of type I and III. Now fixing an $(n + 1)$-line L and counting projective planes and planes of type II on it, we obtain the number of planes of type II on L to be $(n^2 + n + 2)/(n + 2)$ forcing $n + 2|4$ and so $n = 2$ which is impossible. So projective planes do not exist.

We therefore have only three types of planes in S.

We now claim that all planes meet a plane of type II. This is true since any plane of type II has an $(n + 1)$-line L and if $p \in \pi'$, $\langle L,p\rangle \cap \pi'$ is a line meeting L.

As a last major stage in the proof, we show that any two parallel $(n - 1)$-lines M_1 and M_2 of a plane π of type II uniquely determine a parallel class of S. M_1 and M_2 are elements of a parallel class of $n + 1$ $(n - 1)$-lines in π. Let $p \notin \pi$. Let $M_p = \langle M_1,p\rangle \cap \langle M_2,p\rangle$. Suppose M_p is not parallel to some element M_3 of the parallel class on M_1 and M_2 in π. Let $r \in M_3$; let $\langle M_p,r\rangle \cap \pi = M \neq M_3$. Clearly $M\|M_1$ and $M\|M_2$. Let $s \neq r$, $s \in M_3$, and let $M_5 = \langle M_p,s\rangle \cap \pi$. Clearly $M_s\|M_1$, M_2, M, which is impossible in a plane of type II. It follows that each element of the parallel class on M_1 and M_2 in π is parallel to M_p. Let $q \notin \pi$, and define $M_q = \langle M_1,q\rangle \cap \langle M_2,q\rangle$. Then all planes meet π. In particular, $\langle M_p,q\rangle$ meets it in a line M. Since M is in the parallel class of π on M_1 and M_2, we get $M_q \subseteq \langle M_p,q\rangle$ and so $M_p\|M_q$.

If planes of type II do not exist, fix an $(n + 1)$-line L and count v using the $n + 1$ planes of type III on L. We get $v = (n + 1)(n^2 - 1) + n + 1 = n^3 + n^2$. On the other hand, a similar count for v using an $(n - 1)$-

line yields $v = (n + 1)(n^2 - n) + n - 1$, and so we have a contradiction. If some $(n - 1)$-line is not on a plane of type II, we obtain $v = (n + 1)(n^2 - 2n + 1) + n - 1 = n^3 - n^2$ implying that *no* $(n - 1)$-line is on a plane of type II. But since such planes exist, we have another contradiction.

It follows that each $(n - 1)$-line is on a plane of type II. Clearly an $(n - 1)$-line determines at most two parallel classes of S. By the above, it determines precisely two parallel classes of S.

Thus BT holds in S.

If some $(n - 1)$-line is parallel to an n-line, then by the above, *all* $(n - 1)$-lines are parallel to an n-line. In this case, Proposition 7.5.5 implies that BT holds in S, by Kahn's theorem S is embeddable in $PG(3,n)$. Since $(n - 1)$-lines exist, we have $n > 2$. Thus Tallini's theorem now gives the final result. □

7.6 A final embedding theorem

Here we present a major embedding result for d-dimensional linear spaces, $d \geq 3$, due to L. Teirlinck (1986), which in many ways is the analogue of the 2-dimensional results of Metsch presented in Chapter 5 (Theorem 5.3.3). The reader should compare the details of these two important theorems.

In Teirlinck's formulation, only the points, lines and planes play an important role. So it suffices to consider essentially a 3-dimensional linear space. There may indeed exist proper subspaces of S which properly include planes of S, but we impose no restrictions on these entities.

Generalizing above notation, $b_\pi(p)$ denotes the number of lines of π on p.

Theorem 7.6.1. Let S be a 3-dimensional linear space in which each plane π is finite and in which $b_\pi(p) = n_\pi + 1$ is a constant for each $p \in \pi$. Suppose also that for all π and for every line L of π, $k_L - 1 \geq 2(n_\pi + 1)/3$. Then $n = n_\pi$ is independent of π and S is embeddable in a unique way in $PG(d,n)$ for some $d \geq 3$.

PROOF.

Step 1. For distinct coplanar lines L_1 and L_2 and for $p \in \langle L_1,L_2 \rangle = \pi$, $p \notin L_1 \cup L_2$, let $c(p,L_1,L_2)$ be the number of lines on p meeting both L_1 and L_2. Clearly $c(p,L_1,L_2) \geq b_\pi(p) - (b_\pi(p) - k_{L_1}) - (b_\pi(p) - k_{L_2}) = k_{L_1} + k_{L_2} - b_\pi(p) = k_{L_1} + k_{L_2} - n_\pi - 1 \geq k_{L_1} + m_\pi + 1 - n_\pi - 1$, where $m_\pi + 1$ is the smallest line size in π, $\geq k_{L_1} + m_\pi + 1 - (3m_\pi - 2)/2 - 1 = k_{L_1} + 1 - m_\pi/2 \geq k_{L_1} + 1 - (k_{L_1} - 1)/2 = (k_{L_1} + 3)/2$.

Note in particular that this implies $c(p,L_1,L_2) \geq 3$.

Step 2. Here we prove that $n = n_\pi$ is independent of π.

Suppose first of all that π_1 and π_2 are planes which intersect in a line L. Assume that M is a line of π_1 with $M \cap L = \phi$. Let $x \in \pi_1 - (M \cup L)$, $x_1, x_2 \in L$, $x_1 \neq x_2$, $xx_i \cap M = \{x_i'\}$, using step 1. Assume also that $p \in \pi_2 - L$, and $\langle M,p\rangle \cap \pi_2 = \{p\}$. Again using step 1 relative to x, x_1p and $x_1'p$, there are $m \geq 1$ points $p_1, \ldots, p_m$ of $x_1p - \{x_1,p\}$ such that $xp_i \cap x_1'p \neq \phi$ for all i. Put $xp_i \cap x_1'p = \{p_i'\}$. Since $x_2'p_i' \subseteq \langle M,p\rangle$, and $\langle M,p\rangle \cap \pi_2 = \{p\}$, we have $x_2'p_i' \cap x_2p_i = \phi$ for all i. Thus $\langle M,p\rangle$ contains at least m elements (the lines $x_2'p_i'$) of the set $L(x_2) = \{\text{lines } K | K \text{ on } x_2', K \subseteq \langle x,\bar{L}\rangle, \bar{L} \text{ a line of } \pi_2, x_2 \in \bar{L} \neq L, K \cap \bar{L} = \phi\}$ (choose $\bar{L} = x_2p_i$). Similarly using step 1 relative to x, x_2p_1 and $x_2'p_1'$, there are at least $n \geq 1$ points $q_1, \ldots, q_n \in x_2p_1 - \{x_2,p_1\}$ such that $xq_i \cap x_2'p_1' \neq \phi$ for all i. Put $xq_i \cap x_2'p_1', = \{q_i'\}$; and we have $x_1'q_i' \cap x_1q_i = \phi$ since $x_1'q_i' \subseteq \langle M,p\rangle$. So $\langle M,p\rangle$ contains at least n elements of the set $L(x_1) = \{\text{lines } K | K \text{ on } x_1', K \subseteq \langle x,\bar{L}\rangle, \bar{L} \text{ a line of } \pi_2, x_1 \in \bar{L} \neq L, K \cap \bar{L} = \phi\}$.

Since distinct $p \in \pi_2 - L$ with $\langle M,p\rangle \cap \pi_2 = \{p\}$ yield distinct elements of $L(x_1)$ and $L(x_2)$, it follows that the number of such points p is

$$\leq \min\left(\frac{|L(x_2)|}{m(p)}, \frac{|L(x_1)|}{n(p)}\right)$$

where $m(p)$ and $n(p)$ are as m and n above, but considered as functions of p.

Now $|L(x_i)| \leq n_{\langle x,\bar{L}\rangle} + 1 - k_{\bar{L}} \leq \frac{3}{2}(k_{\bar{L}} - 1) - k_{\bar{L}} < \frac{1}{2}(k_{\bar{L}} - 1)$. Thus the minimum above is $<\frac{1}{2}|\pi_2 - L|/\max\{m(p),n(p)\}$. (And the fact that this number is $<\frac{1}{2}|\pi_2 - L|$ will be used in step 4.)

Assume now that $n_{\pi_1} > n_{\pi_2}$. We can then choose $n_{\pi_2} + 2 - k_L$ lines $M_1, \ldots, M_k$ of π_1 on x_1' with $M_i \cap L = \phi$.

By the above, there are at most

$$[n_{\pi_2} + 2 - k_L] \cdot \tfrac{1}{2} \cdot \frac{|\pi_2 - L|}{\max\{m(p),n(p)\}}$$

points $p \in \pi_2 - L$ for which at least one M_i satisfies $\langle M_i,p\rangle \cap \pi_2 = \{p\}$. But this value is

$$\leq \tfrac{1}{3}(n_{\pi_2} + 1) \cdot \tfrac{1}{2} \cdot \frac{|\pi_2 - L|}{\max\{m(p),n(p)\}}$$

$$\leq \tfrac{1}{2}(k_{L^*} - 1) \cdot \tfrac{1}{2} \cdot \frac{|\pi_2 - L|}{\max\{m(p),n(p)\}}$$

for any line L^* of π_2, using our fundamental assumption on n_π.

Now step 1 above implies that $\max\{m(p),n(p)\} \leq \frac{1}{2}(k_{L^*} + 3)$ where L^* is one of the lines x_1p, $x_1'p$, x_2p_1, $x_2'p_1'$. Hence the above number is now

$$\leq \frac{\frac{1}{2}(k_{L^*} - 1) \cdot \frac{1}{2}|\pi_2 - L|}{\frac{1}{2}(k_{L^*} + 3)}$$
$$< |\pi_2 - L|.$$

(Note that in fact, the above is $<\frac{1}{2}|\pi_2 - L|$.)

It follows that there is a point $p \in \pi_2 - L$ such that all $n_{\pi_2} + 2 - k_L$ intersections $\langle M_i,p\rangle \cap \pi_2$ are lines. This gives at least $n_{\pi_2} + 2 - k_L$ lines on p in π_2 missing L, and a contradiction.

Thus $n_{\pi_1} \leq n_{\pi_2}$, and as the roles of π_1 and π_2 are symmetric, we may conclude that $n_{\pi_1} = n_{\pi_2}$.

Now if π_1 and π_2 are planes intersecting in a single point p, choose lines L_i of π_i, $p \in L_i$, $i = 1, 2$. Then by the above, $n_{\pi_1} = n_{\langle L_1,L_2\rangle} = n_{\pi_2}$.

Lastly, if $\pi_1 \cap \pi_2 = \phi$, choose $p \in \pi_1$ and $L \subseteq \pi_2$ and argue as above.

Step 3. (Bundle Theorem.) Let π_1 and π_2 be distinct planes intersecting in the line L. Let L_1 and L_2 be lines of π_1 and π_2 respectively, distinct from L. Suppose there exist points $x_i \in \pi_i - (L \cup L_i)$, $i = 1, 2$. Then by step 1, there are at least three points y of L such that $x_iy \cap L_i \neq \phi$ for $i = 1$ and 2.

We claim also that $k_L \geq 3$ for all lines L. To see this, $n_\pi \geq 1$ for any plane π implies $k_L - 1 \geq 2(n_\pi + 1)/3 \geq 4/3$, implying $k_L - 1 \geq 2$.

Now let L_1, L_2, L_3 and L_4 be disjoint lines of S, no three coplanar, such that L_i and L_j are coplanar for all $i \neq j$, $\{i,j\} \neq \{3,4\}$. We wish to prove that L_3 and L_4 are also coplanar.

By the above, there exists a point $x \in \langle L_2,L_4\rangle - (L_2 \cup L_4)$; and there exist distinct points p_1 and p_2 of L_2 such that for any point $y \in \langle L_1,L_2\rangle - (L_1, \cup L_2)$, $p_ix \cap L_4 = \{q_i\}$ and $p_iy \cap L_1 = \{x_i\}$ are singletons. Choose y in the following way: first, let x_1 be an arbitrary point of L_1, then, let $y \in p_1x_1 - \{p_1,x_1\}$ such that yx meets x_1q_1 in some point u, which is always possible since $c(x,p_1x_1,q_1x_1) \geq 3$ by step 1. Since $\langle u,q_2,x_2\rangle \subseteq \langle x,p_2,y\rangle \cap \langle L_1,L_4\rangle$ we have $x_2 \in q_2u$.

Clearly, if $xy \subseteq \langle L_2,L_3\rangle$ then $\langle x,L_2\rangle = \langle L_2,L_4\rangle = \langle L_2,L_3\rangle$ and so L_3 and L_4 are coplanar. So suppose that $xy \not\subseteq \langle L_2,L_3\rangle$. It follows that on every point of $\langle L_2,L_3\rangle$, there is at most one line of $\langle L_2,L_3\rangle$ coplanar with xy. Let $w \in L_3$ such that p_iw is not coplanar with xy for $i = 1, 2$. By step 1, there are points $w_i \in p_iw - \{p_i,w\}$ such that $xw_i \cap wq_i \neq \phi \neq yw_i \cap wx_i$, $i = 1, 2$.

For $i = 2$, set $W_i = \{w_i$ as above where w varies over L_3, $p_i w$ not coplanar with $xy\}$. Then $|W_i| \geq k_{L_3} - 1$. Also, $(W_1 \cup \{p_1\}) \cap (W_2 \cup \{p_2\}) = \phi$.

We introduce a map on S as follows. For any point $r \neq x$ and such that $xr \cap \langle L_1, L_3 \rangle = \{s\}$ is non-empty, define $\alpha(r) = s$. Let M be any line not coplanar with xy. Denote by M_α the set of points of M for which α is defined. If $M \subseteq \langle L_1, L_3 \rangle$, then clearly $\alpha(M)$ consists of collinear points. Otherwise, $\alpha(M) \subseteq \langle x, M \rangle \cap \langle L_1, L_3 \rangle$ which is contained in a line. So in any case, $\alpha(M)$ is contained in a line. Note that for any point of $W_1 \cup W_2 \cup \{p_1, p_2\}$, the image under α is in $\langle L_1, L_3 \rangle$. Thus, if any line M containing points of $W_1 \cup \{p_1\}$ and of $W_2 \cup \{p_2\}$ had a third point in any of W_1, W_2 or L_3, it would follow that $\alpha(M) \subseteq L_4 \subseteq \langle L_3, q_i \rangle$, $i = 1$ or 2, and so L_3 and L_4 are coplanar. So we may assume that such a line does not exist.

There is a $w_0 \in L_3$, $p_2 w_0$ not coplanar with xy, such that $p_1 \bar{w}_2$ for some $\bar{w}_2 \in W_2$ is coplanar with xy, since otherwise all the lines L_2 and $p_1 w_2$, $w_2 \in W_2$ would be distinct (by the above paragraph), and parallel with L_3 in $\langle L_2, L_3 \rangle$, contradicting the bound on $k_{L_2} - 1$. As $p_1 w_0$ is not coplanar with xy, and $\bar{w}_2 \in p_2 w_0$, we have $\bar{w}_2 \notin xy$, which implies that $p_2 w_0$ is the only line of $\langle L_2, L_3 \rangle$ through $\bar{w}_2$ coplanar with xy. All the lines $\bar{w}_2 w_1$, $w_1 \in W_1$ are then again distinct and parallel with L_3, contradicting the bound on $k_{L_2} - 1$.

It follows that L_3 and L_4 must be coplanar. Thus BT holds in S and applying Kahn's theorem will give us the embedding desired if we can show that X is locally projective.

Step 4. We now prove that S embeds uniquely in a projective space of order $n = n_\pi$, a constant by step 2. It follows from above that $n \geq 2$, and no plane is a near-pencil. It suffices to show that S is locally projective. We do this in two parts. We first of all show that if L and M are coplanar lines, and $p \notin \langle L, M \rangle$, then $\langle L, p \rangle \cap \langle M, p \rangle$ is a line. Let $\pi_1 = \langle L, M \rangle$ and $\pi_2 = \langle p, L \rangle$. Now use the part of the argument of step 2 that refers to this step. By it, there is a line K of π_2 disjoint from L such that $\langle K, M \rangle$ is a plane. If $p \in K$, we are done. Otherwise, let $L_1, \ldots, L_{n+1-k_L}$ be the lines through p missing L in π_2. We now use the part of the counting argument in step 2 where $n_{\pi_1} > n_{\pi_2}$ is assumed, but we may nevertheless still conclude when $n_{\pi_1} = n_{\pi_2} = n$, that there are at most some number $<|\pi_2 - L|$ points satisfying the given conditions. Here we reverse the roles of π_1 and π_2 and conclude using the two occurrences of the inequality $<\frac{1}{2}|\pi_2 - L|$ from Step 2, that there is a point $s \in \pi_1 - L$

such that $\langle K,s\rangle \cap \pi_1$ is a line K' missing L, and all $\langle L_i,s\rangle \cap \pi_1$ are lines L_i' missing L. Since there are precisely $n + 1 - k_L$ lines of π_1 on s missing L, it follows that $K' = L_i'$ for some i. If $L_i' = M$, we are done. If not, M, L_i', K and L_i are lines, no three coplanar, every pair parallel except possibly M and L_i. By step 3, M and L_i are indeed parallel, and we have the first part completed.

We show finally that Pasch's axiom holds in the structure through each point. Let K, L_1, L_2, M_1, M_2 be lines on the point p such that the triples K, L_1, L_2 and K, M_1, M_2 form distinct planes. It suffices to show that the planes $\langle L_1,M_1\rangle$ and $\langle L_2,M_2\rangle$ intersect in a line. Choose $q \in K - p$, $x \in L_1 - p$, $q_2 \in M_2 - p$. We first show that $\langle q,x,q_2\rangle \cap \langle L_1,M_1\rangle$ is a line N_1. If $qq_2 \cap M_1 = \{q_1\}$, then $N_1 = xq_1$. If $qq_2 \cap M_1 = \phi$, then as qq_2 and M_1 are both lines of $\langle M_1,M_2\rangle$, applying the first part of this step gives $\langle q,q_2,x\rangle \cap \langle L_1,M_1\rangle = N_1$, a line. Similarly, $\langle q,q_2,x\rangle \cap \langle L_2,M_2\rangle = N_2$, a line. If $N_1 \cap N_2 = \{y\}$, then $\langle L_1,M_1\rangle \cap \langle L_2,M_2\rangle = py$. If $N_1 \cap N_2 = \phi$, then again using the first part of this step, we get $\langle L_1,M_1\rangle \cap \langle L_2,M_2\rangle = \langle N_1,p\rangle \cap \langle N_2,p\rangle$ is a line. This completes the proof of Theorem 7.6.1.

□

7.7 Exercises

1. Show that if S is a d-dimensional linear space as defined in this chapter, then it has dimension as defined in Section 1.2. Is the converse true?
2. Give an example of a linear space $S = (p, \mathcal{L})$ on which non-isomorphic 3-dimensional structures can be introduced.
3. Suppose S is a 3-dimensional linear space in which $b - v = n^4 + n^2$. Is S necessarily a projective 3-space? If yes, prove it. If no, which other structures satisfy this equality?
4. Let S be a linear space in which each point lies on $n^2 + n + 1$ lines, $n \geq 2$, and such that given any line L and any point p not on L there are precisely n^2 lines of p missing L. Is S a 3-dimensional projective space?
5. Show that all planes of a projective space have the same order.
6. Show that if S is a linear space in which every subspace of dimension 2 is an affine plane, then all these planes have the same order.
7. Find an example of a linear space which is not locally linear at some point.
8. Prove Proposition 7.4.1.
9. Show that the 3-dimensional linear space on 22 points with 2-point

lines and 6-point planes (see the remarks preceding Theorem 7.4.4) does not embed in a projective 3-space of order 4.

10. Prove that the definition of generalized projective geometry given in the proof of Theorem 7.4.4 is equivalent to that given in Section 1.2.
11. Try to find a stronger version of Theorem 7.4.4 by weakening the assumptions as much as possible.
12. Find an example of a non-affine linear space of order 3 in which each plane is affine. Examine carefully Hall's construction.
13. If S is a $\geq$ 2-dimensional linear space in which each plane is a near-pencil, show that there exist precisely two possibilities for S.

7.8 Research problems

1. Find all $\geq$ 2-dimensional linear spaces in which each plane is isomorphic to a fixed linear space π_0 which has precisely two line sizes.
2. Let L_1, L_2, L_3 and L_4 be distinct lines of a projective 3-space, no three on a common point, and not all in a common plane, such that $L_1 \cap L_2 \neq \phi$, $L_2 \cap L_3 \neq \phi$, $L_3 \cap L_4 \neq \phi$ and $L_4 \cap L_1 \neq \phi$. Let S be a linear space having as parameters the parameters of the complement of $L_1 \cup L_2 \cup L_3 \cup L_4$ in the projective space. Is S embeddable in a projective 3-space? Is it the complement of a set of four lines as above?
3. Find all examples of the spaces of Theorem 7.4.3.

7.9 References

Batten, L. M. (1978a), *d*-partition geometries. *Geom. Ded. 7* 63–69.

Batten, L. M. (1978b), Embedding *d*-partition geometries in generalized projective space. *Geom. Ded. 7* 163–174, and Correction (1981), *Geom. Ded. 11* 385–386.

Batten, L. M. (1986), *Combinatorics of Finite Geometries*, Cambridge University Press, Cambridge, New York, Melbourne.

Batten, L. M. (1987a), Locally generalized projective spaces satisfying a bundle theorem. *Geom. Ded. 22* 363–369.

Batten, L. M. (1987b), Locally generalized projective spaces. *J. Geom. 29* 43–49.

Buekenhout, F. (1969), Une caractérisation des espaces affins basée sur la notion de droite. *Math. Z.* 111, 367–371.

Buekenhout, F. and Deherder, R. (1971), Espaces linéaires finis à plans isomorphes. *Bull. Soc. Math. Belg.* 348–359.

Ceccherini, P. V. and Tallini, G. (1986), A new class of planar π-spaces and some related topics: (n, d)-systems and (σ, n)-spaces. *J. of Geom. 27* 69–86.

Delandtsheer, A. (1982), Finite planar spaces with isomorphic planes. *Discrete Math. 42* 161–176.

Delandtsheer, A. (1983), Some finite rank 3 geometries with two isomorphic linear residues. *Archiv Math 41* 475–477.

Dembowski, P. (1968), *Finite Geometries*. Springer-Verlag, New York.

Doyen, J. and Hubaut, X. (1971), Finite regular locally projective spaces. *Math. Z. 119* 83–88.

Frank, R. (1985), A proof of the bundle theorem for certain semi-modular locally projective lattices of rank 4. *J. Comb. Theory* (A) *39* 222–225.

Hafner, H. P. (1993), *Eingeschraenkte planare Raüme*, Ph.d thesis, University of Mainz.

Hall, Jr., M. (1960), Automorphisms of Steiner triple systems. *IBM J. Res. Develop. 4* 460–472.

Herzer, A. (1976), Projektiv darstellbare stark planare Geometrien vom Rang 4. *Geom. Ded. 5* 467–484.

Herzer, A. (1979), Büschelsätze zur Charakterisierung projektiv darstellbarer Zykelebenen. *Math. Z. 164* 215–238.

Herzer, A. (1981), Semimodular locally projective lattices of rank 4 from Von Staudt's point of view. *NATO Advanced Study Institute*, Series C. Reidel, Boston, London 373–400.

Hirschfeld, J. W. P. (1979), *Projective geometries over finite fields*. Clarendon Press, Oxford, New York.

Kahn, J. (1980a), Locally projective-planar lattices which satisfy the bundle theorem. *Math. Z. 175* 219–247.

Kahn, J. (1980b), Inversive planes satisfying the bundle theorem. *J. Comb. Theory* (A) *29* 1–19.

Kahn, J. (1982), Finite inversive planes satisfying the bundle theorem. *Geom. Ded. 12* 171–187.

Kantor, W. M. (1974), Dimension and embedding theorems for geometric lattices. *J. Comb. Theory* (A) *17* 173–195.

Karzel, H. and Pieper, I. (1970), Bericht über geschlitzte Inzidenzgruppen. *J.-ber. Deutsch. Math.-Verein.* 72 70–114.

Leonard, D. A. (1982), Finite π-spaces. *Geom. Ded. 12* 215–218.

Metsch, K. (1993), The theorem of Totten for planar spaces. *Rendi. Roma* (to appear).

Sörensen, K. (1988), Ein Beweis von H. Karzel und I. Pieper. *J. Geom. 32* 131–132.

Tallini, G. (1956), Sulle K-calotte di uno spazio lineare finito. *Ann. Math. 42* 119–164.

Teirlinck, L. (1975), On linear spaces in which every plane is either projective or affine. *Geom. Ded. 4* 39–44.

Teirlinck, L. (1986), Combinatorial properties of planar spaces and embeddability. *J. Comb. Theory* (A) *43* 291–302.

Wille, R. (1967), Verbandstheoretische Charakterisierung *n*-stufiger Geometrien. *Arch. Math.* 18 465–468.

Wille, R. (1971), On incidence geometries of grade *n*. *Atti Conv. Geom. Comb. Appl.* Perugia 421–426.

Wyler, O. (1953), Incidence geometry. *Duke Math. J. 20* 601–610.

8
Group action on linear spaces

8.1 Reasons for this chapter and some background

Up to this point in the book, the results we have presented are aimed at finding relationships between linear spaces and projective spaces. In this chapter we consider the broader question of how linear spaces relate to each other. This is too general a problem to tackle with our earlier methods. Our approach will be to introduce groups. This is a fairly new approach, with relevant articles first appearing perhaps twenty years ago. A large part of the research has been aided by recent important developments in group theory – in particular, the classification of finite simple groups. We shall list in this section those group-theoretic results needed for the chapter. We give them without proof, but of course refer to articles or books in which proofs may be found.

We emphasize the fact that there are many results about groups operating on linear spaces, which we shall *not* include here, because the results are about the group structure rather than the linear space structure.

The first major result, to which we shall refer later, is the classification of all finite 2-transitive groups. This classification is a consequence of the classification of all finite simple groups. For the latter, we refer the reader to D. Gorenstein (1982). The result we give here can be found in W. M. Kantor (1985).

Theorem 8.1.1. Let G be a 2-transitive group of permutations of a finite set X on v elements. Then the only possibilities for G are as follows:

(A) G has a simple normal subgroup N, and $N \subseteq G \subseteq Aut(N)$, where N and v are given below:

(1) A_v, $v \geq 5$, the alternating group on v elements.

(2) $PSL(d,q)$, $d \geq 2$, $v = (q^d - 1)/(q - 1)$ (two representations if $d > 2$); here (d,q), $\neq (2,2)$, $(2,3)$.
(3) $PSU(3,q)$, $v = q^3 + 1$, $q > 2$.
(4) $Sz(q)$, $v = q^2 + 1$, $q = 2^{2m+1} > 2$.
(5) ${}^2G_2(q)'$, $v = q^3 + 1$, $q = 3^{2m+1}$.
(6) $Sp(2n,2)$, $n \geq 3$, $v = 2^{2n-1} \pm 2^{n-1}$.
(7) $PSL(2,11)$, $v = 11$ (two representations).
(8) Mathieu groups M_v, $v = 11,12,22,23,24$ (two representations for M_{12}).
(9) M_{11}, $v = 12$.
(10) A_7, $v = 15$ (two representations).
(11) HS (Higman–Sims group), $v = 176$ (two representations).
(12) .3 (Conway's smallest group), $v = 276$.

(B) G has a regular normal subgroup N which is elementary abelian of order $v = p^d$, where p is a prime. Identify G with a group of affine transformations $x \to x^g + c$ of $X = GF(p^d)$, where $g \in G_0$, the stabilizer of 0. Then one of the following occurs:

(1) $G \subseteq A\Gamma L(1,v)$.
(2) $SL(n,q) \lhd G_0$, $q^n = p^d$.
(3) $S_p(n,q) \lhd G_0$, $q^n = p^d$.
(4) $G_2(q)' \lhd G_0$, $q^6 = p^d$, q even.
(5) $G_0 \simeq A_6$, or A_7, $v = 2^4$.
(6) $SL(2,3)$ or $SL(2,5) \lhd G_0$, $v = p^2$, $p = 5,7,11,19,23,29$ or 59 or $v = 3^4$.
(7) G_0 has a normal extraspecial subgroup E of order 2^5, and G_0/E is isomorphic to a subgroup of S_5, where $v = 3^4$.
(8) $G_0 = SL(2,13)$, $v = 3^6$.

We shall also be using the next result.

Theorem 8.1.2. (T. G. Ostrom and A. Wagner 1959.) Let S be a finite projective plane whose automorphism group is 2-transitive on points. Then S is Desarguesian.

8.2 Point and line orbits

Let $S = (p,\mathcal{L})$ be an arbitrary linear space. *The automorphism group* of S is the set of all permutations (one-to-one, onto maps) of the point set and of the line set of S which preserve point-line incidence (and therefore also point-line non-incidence).

We use *Aut*(S) to denote this group.

The *point orbits of S under G*, G a subgroup of Aut (S), are the equivalence classes of p under the relation R defined by: pRq if and only if there exists $g \in G$ such that $g(p) = q$.

The *line orbits of S under G* are similarly defined.

As we are also interested in relationships between points and lines, we generalize the above ideas to ordered and unordered pairs: Let (a,b) be an ordered (or unordered) pair of elements a, $b \in p \cup \mathcal{L}$. Let X be any subset of the set of all such ordered (or of all such unordered) pairs. The *orbits of X* are the equivalence classes of X under the relation R defined by $(a,b)R(a',b')$ if and only if there exists $g \in G$ such that $g(a) = a'$ and $g(b) = b'$ (or such that $\{g(a),g(b)\} = \{a',b'\}$ for unordered pairs).

The ideas here can clearly be extended to ordered and unordered k-tuples, but we shall not consider them here.

If $p \in L$, the pair (p,L) is called a *flag*; if $p \notin L$, (p,L) is called an *antiflag*.

We say that $G \subseteq$ Aut(S) is *point*, *line*, *flag* or *antiflag transitive* if there is a single point, line, flag or antiflag orbit respectively, under the relation R above. If G = Aut(S), we may also say abusively that S is transitive.

Example 8.2.1. (Buekenhout 1968.) Let S be the unit disc centered at the origin in the Euclidean plane. That is, the points (x,y) of S are all pairs of real numbers x and y satisfying $x^2 + y^2 \leq 1$, and the lines of S are the restrictions of the usual lines of the Euclidean plane to the above point set, as long as such restrictions have at least two points.

It is not too difficult to check that S is line transitive. The standard Euclidean translations, reflections, dilatations and rotations can be used. See, for example, L. M. Batten (1986).

S has two point orbits under Aut(S), on the other hand. To see this, let p be an interior point and p' be a boundary point, and suppose that $g \in$ Aut(S) maps p to p'. It is possible to find lines L_1 and L_2 not on p such that each line on p meets either L_1 or L_2. Consider $g(L_1)$ and $g(L_2)$. Since any line on p' is the image under g of some line on p, it follows that any line on p' meets $g(L_1)$ or $g(L_2)$. However, it is always possible to find a line L on p' meeting neither $g(L_1)$ nor $g(L_2)$. Thus, point-line incidence is not preserved, and g cannot be in Aut(S).

The number of point and of line orbits of a linear space S under a subgroup G of Aut(S) will be denoted by v_G and b_G respectively. Thus, in the above example, $v_{\text{Aut}(S)} = 2$ and $b_{\text{Aut}(S)} = 1$. In fact, we show next that this situation never occurs in the finite case. The proof we give is due to R. E. Block (1967) for designs, but we have specialized it to the linear space situation.

Theorem 8.2.2. Let S be a non-trivial finite linear space, and G a subgroup of Aut(S). Then $v_G \leq b_G$.

PROOF. Let $P_1, \ldots, P_{v_G}$ be the point orbits and $L_1, \ldots, L_{b_G}$ the line orbits of S under G, and consider the $v \times b$ incidence matrix M of S where the rows/columns are the points/lines grouped according to their orbits. Since S is a linear space, the rows of this matrix are independent. (See Exercise 1.8.2.) Consider now the $|P_i| \times |L_j|$ matrix M_{ij} formed by taking the entries from M in the rows and columns corresponding to P_i and L_j respectively, $l \leq i \leq v_G$, $1 \leq j \leq b_G$. In each M_{ij} the row sums are constant, as are the column sums. Let M_{cs} be the $v_G \times b_G$ matrix whose (i,j)-th entry is the constant column sum s_{ij} from M_{ij}. Now in M_{cs}, the rows must be independent, as a dependence relation here would give a dependence relation with the same, but perhaps repeated, coefficients in M. However, the rank of M_{cs} is at most b_G, and so $v_G \leq b_G$. □

Note the very pretty analogy between Theorem 8.2.2 and Theorem 1.5.5, the Fundamental Theorem. It would be nice to extend this analogy even further by getting a simple classification when $v_G = b_G$. This seems to be far from trivial. J. Saxl (1981, 1983), however, has proved that for Steiner triple systems on at least eight points, $v_G = b_G$ implies $v_G \leq 3$. For projective planes, the corollary below is helpful, but we refer the reader also to the research problem section at the conclusion of the chapter.

Corollary 8.2.3. Let S be a finite linear space in which $v = b$, and G be a subgroup of Aut(S). Then $v_G = b_G$.

PROOF. In the proof of Theorem 8.2.2, it suffices to use columns throughout, instead of rows, since the set of $b = v$ matrix columns is linearly independent. □

The next theorem is a recent result giving information on the possible pairs (v_G, b_G). This same result is *not* true for the pairs (v,b).

Theorem 8.2.4. (A. Blokhuis, A. Brouwer, A. Delandtsheer, J. Doyen 1987.) For all integer pairs (v_G, b_G) with $v_G \le b_G$, there is a finite linear space with v_G point orbits and b_G line orbits.

PROOF. We construct such a space.

For any $n \ge 2$, begin with a projective geometry $PG(n, q^{2^{n-1}})$ where q is a prime power. Write $q_i = q^{2^{i-1}}$, $1 \le i \le n$.

Let $p = \{[x_1, \ldots, x_n, 1] | x_i \in GF(q_i)\}$.

Let $\mathcal{L}$ = {lines of $PG(n, q_n)$ restricted to points of p and having at least two points in p}.

The system $S = (p, \mathcal{L})$ is a linear space. We proceed to calculate the line sizes.

For distinct points $[x_1, \ldots, x_n, 1]$ and $[y_1, \ldots, y_n, 1]$ of p, we can generate all other points on the corresponding line by considering

$$\lambda[x_1, \ldots, x_n, 1] + [y_1, \ldots, y_n, 1] \qquad \forall\, \lambda \in GF(q_j),\ \lambda \ne -1.$$

Then

$$\begin{gathered}
[\lambda x_1 + y_1, \ldots, \lambda x_n + y_n, \lambda + 1] \in p \text{ iff} \\
(\lambda + 1)^{-1} (\lambda x_i + y_i) \in GF(q_i) \qquad \forall\, i \text{ iff} \\
(\lambda + 1)^{-1} [(\lambda + 1)x_i + y_i - x_i] \in GF(q_i) \qquad \forall\, i \text{ iff} \\
x_i + (\lambda + 1)^{-1} (y_i - x_i) \in GF(q_i) \qquad \forall\, i \text{ iff} \\
(\lambda + 1)^{-1} (y_i - x_i) \in GF(q_i) \qquad \forall\, i \text{ iff} \\
(\lambda + 1)^{-1} \in GF(q_i) \quad \text{or} \quad x_i = y_i \qquad \forall\, i \text{ iff} \\
\lambda \in GF(q_i) \quad \text{or} \quad x_i = y_i \qquad \forall\, i.
\end{gathered}$$

Hence, if $x_i = y_i$, $1 \le i \le j - 1$, but $x_j \ne y_j$, then $\lambda \in GF(q_j)$, $\lambda \ne -1$. So the line sizes in S are q_i, $1 \le i \le n$.

Then we extend S to a new linear space $S^* = (p^*, \mathcal{L}^*)$ by adding the points corresponding to $\lambda = -1$: $[x_1, \ldots, x_n, 1] - [y_1, \ldots, y_n, 1]$ to p, and by adding one new line, p_∞, consisting of all these new points, to $\mathcal{L}$. The line sizes in S^* are now $q_i + 1$, $1 \le i \le n$, and $|p_\infty|$.

Let $p \in p^*$, $p \notin p_\infty$. All lines on p meet p_∞. If $p = [x_1, \ldots, x_n, 1]$, the lines on p of size $q_i + 1$ are generated by points $[y_1, \ldots, y_n, 1]$ where $x_1 = y_1, \ldots, x_{i-1} = y_{i-1}$, $x_i \ne y_i$. Here p and the points of p_∞ are not included, so there are $(q_i - 1)(q_{i+1} \ldots q_n)/(q_i - 1) = q_{i+1} \ldots q_n$ $(q_i + 1)$-lines on p in S^*. Therefore $|p_\infty| = 1 + q_n + \ldots + (q_2 \ldots q_n)$.

Since there are $n + 1$ line sizes in S^*, it follows that there are at least $n + 1$ line orbits in Aut (S^*).

All points not on p_∞ have the same size, $|p_\infty|$, as above. Consider a point $p \in p_\infty$. Let L and L' be distinct lines on p, and on the points $[x_1,\ldots,x_n,1]$, $[y_1,\ldots,y_n,1]$ and $[x'_1,\ldots,x'_n,1]$, $[y'_1,\ldots,y'_n,1]$ respectively, such that $p = [x_1,\ldots,x_n,1] - [y_1,\ldots,y_n,1] = [x'_1,\ldots,x'_n,1] - [y'_1,\ldots,y'_n,1]$. Suppose $|L| = q_s + 1$ and $|L'| = q_t + 1$. Then $p = [0,\ldots,0, x_s - y_s \neq 0,\ldots,0] = [0,\ldots,0,x'_t - y'_t \neq 0,\ldots,0]$, implying $s = t$. Thus a point of p_∞ is on lines of only one size. Hence there exist at least $n + 1$ point orbits of Aut (S^*).

But now consider the group of projectivities generated by the matrices

$$\begin{pmatrix} a_{11} & 0 \ldots & 0 & b_1 \\ a_{21} & a_{22}\, 0 \ldots & 0 & b_2 \\ \vdots & & \vdots & \vdots \\ a_{n1} & \ldots & a_{nn} & b_n \\ 0 & \ldots & 0 & 1 \end{pmatrix} \qquad \begin{matrix} a_{ii} \neq 0 \\ \\ \\ a_{ij}, b_i \in GF(q_i). \end{matrix}$$

This subgroup of Aut (S^*) is transitive on p and on lines of the same size, and so has precisely $n + 1$ point and line orbits. It follows that Aut (S^*) itself has precisely $n + 1$ point and line orbits. Thus $(n + 1, n + 1)$ is 'realizable' for all $n \geq 2$.

Removing r orbits from p_∞, $1 \leq r \leq n - 1$, or n orbits and the line p_∞ we see that the pairs $(n + 1, n + 1 - r)$, $r \leq n - 1$ and $(1,n)$ are also realizable. This leaves only the pairs $(1,1)$ and $(2,2)$; examples of these were mentioned above. □

8.3 Relationships between transitivities

In Section 8.2, we defined transitivity relative to sets of singletons (points or lines), and sets of pairs (flags or antiflags). This idea can of course be generalized to sets of k-tuples of elements of $p \cup \mathcal{L}$. (One might find some interesting problems about transitivity on 3-tuples of points, where the points are not collinear, i.e., form a triangle.)

If X is a subset of the set of all *ordered* k-tuples of $p \cup \mathcal{L}$, we say that *G is transitive on X*, G a subgroup of Aut(S), if there is a single orbit on X under the relation: $(x_1,\ldots,x_k)R(y_1,\ldots,y_k)$ if and only if there is a $g \in G$ such that $g(x_i) = y_i$ for all i.

If X is a subset of the set of all *unordered* k-tuples of $p \cup \mathcal{L}$, we say that G is *homogeneous* on X, G a subgroup of Aut(S), if there is a single orbit on X under the relation $(x_1,\ldots,x_k)R(y_1,\ldots,y_k)$ if and only if for some $g \in G$, $\{g(x_1),\ldots,g(x_k)\} = \{y_1,\ldots,y_k\}$.

If $G = \mathrm{Aut}(S)$, we may also, abusively, say that *S is transitive/homogeneous on X*.

If X is the set of *all* ordered k-tuples of points (lines) and G is transitive on X for some subgroup G of Aut(S), we say that S is *k-transitive on points (lines)*. A similar definition exists for *k-homogeneous*. Clearly, k-transitivity on points (lines) implies k-homogeneity on points (lines).

Since the point-line structure of S is of primary interest to us here, we restrict ourselves to the cases $k = 1$ and 2. For these cases, it is easy to see that only the following sets X are of any interest:

(i) $X = \{\{p\}|p \in \mathcal{P}\}$.

(ii) $X = \{\{L\}|L \in \mathcal{L}\}$.

(iii) $X = \{(p_1,p_2)|p_i \in \mathcal{P}\}$ or $\{\{p_1,p_2\}|p_i \in \mathcal{P}\}$.

(iv) $X = \{(L_1,L_2)|L_i \in \mathcal{L}\}$ or $\{\{L_1,L_2\}|L_i \in \mathcal{L}\}$.

(v) $X = \{(p,L)|p \in \mathcal{P}, L \in \mathcal{L}, p \in L\}$.

(vi) $X = \{(p,L)|p \in \mathcal{P}, L \in \mathcal{L}, p \notin L\}$.

(vii) $X = \{(L_1,L_2)|L_i \in \mathcal{L}, L_1 \cap L_2 = \phi\}$ or $\{\{L_1,L_2\}|L_i \in \mathcal{L}, L_1 \cap L_2 = \phi\}$.

(viii) $X = \{(L_1,L_2)|L_i \in \mathcal{L}, L_1 \cap L_2 \neq \phi\}$ or $\{\{L_1,L_2\}|L_i \in \mathcal{L}, L_1 \cap L_2 \neq \phi\}$.

We now list some of the relationships between the transitivities and homogeneities represented by the sets in (i) through (viii).

1. Line transitivity implies point transitivity.

This follows from Block's Theorem 8.2.2. That the converse is false can be seen easily by taking the example of the projective plane of order 2 with one point deleted.

2. If S is 2-transitive or 2-homogeneous on points, then it is line transitive.

This result is trivial. Note that if S is transitive on lines, then all line sizes are equal and hence all point sizes are equal. Thus S is a 2-$(v,k,1)$ design, or in other notation a Steiner system $S(2,k,v)$. In this case, we shall also use the latter to denote the constant number of lines on a point.

3. If S is flag or antiflag transitive, then S is line transitive.

This result is also trivial.

The next result is a partial converse to result (3), and is due to A. R. Camina and T. M. Gagen (1984). We give it as a theorem.

4. *Theorem 8.3.1. If S is a 2-$(v,k,1)$ design with $k|v$, and if S is line transitive, then S is flag transitive.*

PROOF. Since $G = \mathrm{Aut}(S)$ is line transitive, it suffices to show that for any line L, G_L, the stabilizer of L, is transitive on the points of L.

Let p and q be points of L. Since S is line transitive, it is point transitive by result (1). Thus for each point in S there is a $g \in G$ mapping p to each of the v points of S. Therefore $[G\colon G_p] = v$, where G_p is the stabilizer of p in G. Similarly, $[G\colon G_L] = b$, the number of lines of S. But $b = v(v - 1)/k(k - 1)$. From $[G\colon G_{p,q}] = [G\colon G_p][G_p\colon G_{p,q}] = [G\colon G_L][G_L\colon G_{p,q}]$ (where $G_{p,q}$ is the stabilizer of p and of q in G), we obtain $[G_p\colon G_{p,q}] = (v - 1)/k(k - 1)\,[G_L\colon G_{p,q}]$. Since $k|v$, k and $v - 1$ are coprime. Therefore $k|[G_L\colon G_{p,q}]$.

Let n be a prime divisor of k, and n^a the highest power of n dividing k. The above shows that $n^a|[G_L\colon G_{p,q}]$ for all pairs of points p,q of L. Suppose the point p of L is in an orbit Γ_p of G_L whose size is not divisible by n^a. Since $[G_L\colon G_{p,q}] = [G_L\colon G_p][G_p\colon G_{p,q}] = |\Gamma_p|[G_p\colon G_{p,q}]$, it follows that $n|[G_p\colon G_{p,q}]$ for any $q \neq p$ in L. Let Γ be the union of orbits of L under G_L whose order is not divisible by n^a. If $q \in \Gamma_p$, then $|\Gamma_p| \equiv 1 \pmod{n}$; while if $q \in \Gamma - \Gamma_p$, we have $|\Gamma - \Gamma_p| \equiv 0 \pmod{n}$. It follows that $|\Gamma| \not\equiv 0 \pmod{n}$, whereas this is clearly not the case. So n^a divides the order of each orbit of G_L over L. Since this is true for all prime divisors of k, we conclude that G_L has a single point orbit on L. □

5. If S is 2-transitive on points, then it is flag transitive.

This result is easy.

6. *Theorem 8.3.2. If S is antiflag transitive, then S is 2-transitive on points.*

PROOF. Let p be a point of S and let G_p be the stabilizer of p in S under G, a subgroup of $\mathrm{Aut}(S)$ antiflag transitive on S. Let n be the number of point orbits of $S - p$ under G_p, and m the number of line orbits containing p, under G_p. The number of point orbits of S under G_p is then $n + 1$; and since G is antiflag transitive on S, the number of line orbits of S

under G_p is $m + 1$. By Theorem 8.2.2, $n + 1 \leq m + 1$. It follows that $n = m$, and in any line orbit of the lines on p under G_p, all the points distinct from p lie in the same point orbit.

Now fix pairs (p,q) and (x,y), $p \neq q$, $x \neq y$. Then there is a map of G taking pq to xy by result (3), with $p \to p'$ and $q \to q'$, say. By the above paragraph, there is a map of $G_{p'}$ taking q' to y, and a map of G_y taking p' to x. The composition of all three maps takes (p,q) to (x,y) under G. □

Results (5) and (6) together yield the rather surprising fact that if S is antiflag transitive then it is flag transitive. The original proof, at least for projective planes, seems to be due to T. G. Ostrom (1958). See also W. M. Kantor (1972) and A. Delandtsheer (1984a).

The next three properties are easy to see.

7. If S is homogeneous on pairs of intersecting lines, then S is transitive on points.
8. If S is 2-homogeneous on lines, then S is homogeneous on pairs of (non-)intersecting lines.
9. If S is 2-transitive on lines, then S is 2-homogeneous on lines and is transitive on pairs of (non-)intersecting lines.

10. *Theorem 8.3.3. (Delandtsheer 1984b.) If S is transitive both on pairs of intersecting lines and on pairs of non-intersecting lines, then S is 2-transitive on points.*

PROOF. The fact that $G = \text{Aut}(S)$ is transitive on pairs of intersecting lines implies that G is transitive on flags. Fix a line L. The assumptions of the theorem imply that the stabilizer G_L of L has exactly three orbits on the lines of S. By Theorem 8.2.2, it has at most three orbits on the points of S.

Clearly G_L cannot have a single point orbit on S. (S is non-trivial!) Suppose G_L has precisely two point orbits on S. The two assumptions of the theorem and the fact that G is consequently flag transitive now yield the result that G has exactly two orbits on the set $\{(p,L) | p \text{ a point}, L \text{ a line}\}$. Thus for any point p, G_p, the stabilizer of p, has exactly two orbits on the lines of S. Applying Theorem 8.2.2 once again gives the conclusion that G_p has at most two orbits on the points of S. But G is easily seen to be point transitive and therefore now, 2-transitive on points.

Suppose then that G is not 2-transitive on points. Then G_L has precisely three point orbits. In fact, G_L has exactly two point orbits, say O and O' on $S - L$. The conditions of the theorem then imply that all lines meeting

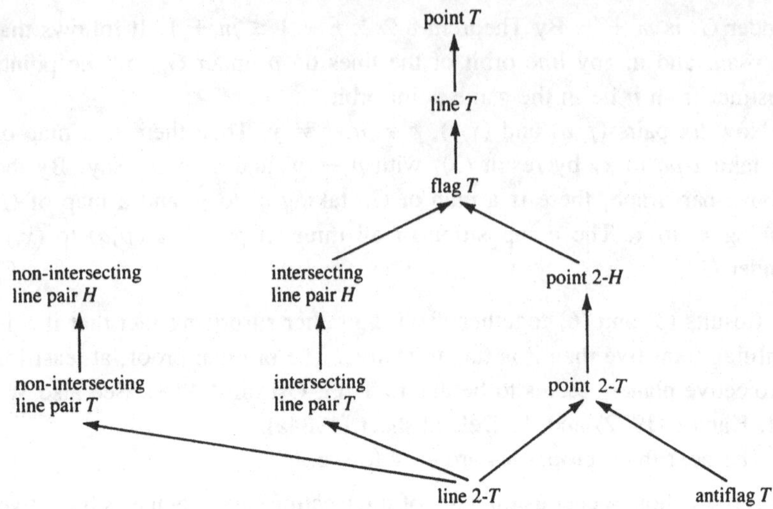

Figure 8.3.1.

L have a constant number c of points of O, and that all lines missing L have a constant number d of points of O. Letting $p \in O$ and $p' \in O'$ and counting the pairs (q,pq), and the pairs $(q,p'q)$ where $q \in O$, $q \neq p$, we get $|O| - 1 = (r - k)(d - 1) + k(c - 1)$, and $|O| = (r - k)d + kc$. But this gives us $r = 1$ and a contradiction. □

A recent paper of F. Buekenhout, A. Delandtsheer and J. Doyen (1988) analyzes the relationships between transitivities and primitivities on linear spaces, and gives a diagram which represents these relationships in a compact way. We mimic this idea of a diagram to abbreviate and clarify the results of this section in Figure 8.3.1. The letters T and H stand for transitive and homogeneous respectively.

8.4 Characterization theorems

Before stating and proving the theorems of this section, we need to describe some of the exceptional linear spaces which arise.

A *unital* is a design with parameters of the form $v = s^3 + 1$, $k = s + 1$, $b = s^2(s^2 - s + 1)$, $r = s^2$, $\lambda = 1$.

The absolute points and non-absolute lines of a Hermitian (also known as unitary) polarity in $PG(2,s^2)$ (see, for instance, L. M. Batten 1986) form such a design, called a *Hermitian unital*.

Let G be a Ree group (R. Ree 1961a,b). G has order $s^3(s^3 + 1)(s - 1)$

where $s = 3^{2m+1}$, $m > 0$. If P denotes the normalizer of a Sylow 3-subgroup of G, then P is maximal in G of index $s^3 + 1$. So the permutation representation Γ of G on the set p of right cosets of P is primitive of degree $s^3 + 1$. J. Tits (1960) showed that Γ is 2-transitive on p, that the stabilizer $\Gamma_{x,y}$, $x, y \in p$, of order $s - 1$ is cyclic and therefore contains exactly one involution, and that the total number of fixed points of this involution is $s + 1$. Taking p as the set of points, and for each pair $x, y \in p$, letting the $s + 1$ fixed points described above be the *line* on x and y, we get a linear space with the parameters of a unital. These unitals were discovered by H. Lüneburg (1966), and are called the *Ree unitals*. They are not isomorphic to the Hermitian unitals. We shall also need the fact that the stabilizer $\Gamma_{x,y,z}$ of three collinear points has order 2.

A *nearfield* (P. Dembowski 1968) is a triple $(N, +, \cdot)$ where N is a set and $+$ and $\cdot$ are binary operations on N, such that $(N, +)$ is a group with identity 0 and $(N - 0, \cdot)$ is a group. We require in addition that $x \cdot 0 = 0$ for all $x \in N$ and $(x + y) \cdot z = x \cdot z + y \cdot z$ for all $x, y, z \in N$.

Such an algebraic structure can be used to coordinate an affine (and projective) space. An affine plane constructed over a nearfield will be called a *nearfield affine plane*. We note for future reference that there is a unique nearfield affine plane of order 9.

Let $V = V(6,3)$ be the 6-dimensional vector space over $GF(3)$ and let U be the subspace of V defined by the basis vectors $(1,0,0,0,0,0)$, $(0,1,0,0,0,0)$ and $(0,0,1,0,0,0)$. The subgroup G of $GL(6,3)$ generated by

$$s = \begin{pmatrix} 0 & 2 & 0 & 0 & 0 & 0 \\ 2 & 0 & 2 & 0 & 0 & 0 \\ 1 & 0 & 0 & 0 & 0 & 0 \\ 0 & 0 & 0 & 2 & 0 & 1 \\ 0 & 0 & 0 & 2 & 0 & 0 \\ 0 & 0 & 0 & 0 & 2 & 0 \end{pmatrix} \qquad h = \begin{pmatrix} 0 & 0 & 0 & 2 & 0 & 0 \\ 0 & 0 & 0 & 2 & 2 & 0 \\ 0 & 0 & 0 & 0 & 1 & 2 \\ 1 & 0 & 0 & 0 & 0 & 0 \\ 1 & 1 & 0 & 0 & 0 & 0 \\ 2 & 2 & 1 & 0 & 0 & 0 \end{pmatrix}$$

$$r = \begin{pmatrix} 2 & 0 & 0 & 2 & 0 & 0 \\ 2 & 1 & 0 & 2 & 1 & 0 \\ 1 & 2 & 2 & 1 & 2 & 2 \\ 2 & 0 & 0 & 1 & 0 & 0 \\ 2 & 1 & 0 & 1 & 2 & 0 \\ 2 & 2 & 2 & 2 & 1 & 1 \end{pmatrix}$$

is isomorphic to $SL(2,13)$ and is transitive on the non-zero vectors of V. An affine plane of order 27 can now be constructed. The points are the vectors of V. The lines are the images of U under G and under the translation group of V. The construction is due to C. Hering (1969), and we call the plane the *Hering affine plane*.

Hering (1985) also constructed two $S(2,9,9^3)$s each with automorphism group the same sporadic doubly transitive automorphism group of degree 3^6, which is the automorphism group of the Hering affine plane. The two Steiner systems are constructed as follows. Let $V = V(6,3)$ and G be as above. G contains subgroups X_1 and X_2 of order 24 such that if I is the identity matrix, then $-I \in X_1$ and $X_1/\langle -I\rangle$ is a dihedral group of order 12, while $X_2 \approx SL(2,3)$. Let $i \in \{1,2\}$. The group X_i leaves invariant a subspace U_i of dimension 2 of V, and X_i is the stabilizer of G on U_i, as X_i is maximal in G, while G acts irreducibly on V. Therefore, the number of images of U_i under G is 91. Since G is transitive on $V - 0$, every non-zero vector must be contained in one of these images. This implies that U_i^G is a partition of V. Let $L_i = \{v + U_i^g | v \in V, g \in G\}$, $i = 1,2$. Then each (V,L_i) is a linear space with 3^6 points and 3^2 points per line.

We now define a *Netto system*. (The name *Netto system* seems to be due to P. Dembowski (1968, p. 98), but R. M. Robinson (1975) notes that they are different from the systems constructed by Netto in 1893.) Let $q = p^n$, $p \equiv 7 \pmod{12}$ prime, n odd. Since $q - 1 \equiv 0 \pmod 6$, there are exactly two primitive sixth roots of unity ϵ_1, ϵ_2 in $F = GF(q)$. Note that ϵ_1, ϵ_2 are non-squares and are the roots of $x^2 - x + 1 = 0$, so that $\epsilon_1 + \epsilon_2 = \epsilon_1\epsilon_2 = 1$. Define a relation $<$ on F by $u < v$ if and only if $v - u$ is a non-zero square in F. As -1 is not a square, we have either $u < v$ or $v < u$ for all $u \neq v$ in F. Let the points of the Netto system be the elements of F. For $u < v$, let $w = f(u,v) = u\epsilon_1 + v\epsilon_2$. It is easy to check that if $v < w$ then $f(v,w) = u$ and if $w < u$ then $f(w,u) = v$. Define the lines to be all sets $\{u,v,f(u,v)\}$ where $u < v$. It follows that this is an $S(2,3,q)$. If $q = 7$, we have the Fano plane. This is the only value for q leading to a projective plane.

An *inversive plane* is a pair (p,c) where p is a set of $n^2 + 1$ points and c is a set of $n(n^2 + 1)$ *circles* each having $n + 1$ points, such that any three points are on a unique circle. It follows that each point is on $n(n + 1)$ circles. A linear space structure can be introduced by defining lines as pairs of points. By a *4-chain* we mean any quadruple $C_1, \ldots, C_4$ of circles such that no three of the C_i have a common point, but $C_i \cap C_{i+1} \neq \phi$, with the subscripts taken modulo 4. Put $C_i \cap C_{i+1} = \{a_i,b_i\}$, where a_i may equal b_i. We say that an inversive plane is *Miquelian* if it satisfies the following condition: For any 4-chain $C_1, \ldots, C_4$, the points $a_1, \ldots, a_4$ are on a common circle if and only if the points $b_1, \ldots, b_4$ are on a common circle.

We now move to the actual theorems.

Theorem 8.4.1. (Kantor 1985.) If S is a finite linear space with automorphism group 2-transitive on points, then S is one of:

(a) a Desarguesian affine or projective space (in the latter case, two points per line is allowed),
(b) a Hermitian or Ree unital,
(c) the Hering affine plane of order 27 or the nearfield affine plane of order 9,
(d) one of the two Steiner systems $S(2,9,9^3)$ due to Hering and described above.

PROOF. Since the proof relies heavily on group theory, we merely indicate the method here. It is based on the classification of all 2-transitive groups given in Theorem 8.1.1; these are considered one by one. All possibilities are excluded except the following: $G = PSL(d,q)$, $d \geq 3$ yields $S = PG(d-1,q)$; $G = PSU(3,q)$, $v = q^3 + 1$ yields the Hermitian unital; $G = {}^2G_2(q)$ yields the Ree unital; $G = A_7$, $v = 15$ yields $S = PG(3,2)$; $G \subseteq A\Gamma L(1,v)$ yields $S = AG(d,q)$; if $SL(n,q)$, $S_p(n,q)$ or $G_2(q)'$ $\lhd G_0$, then $S = AG(d,q)$; if $SL(2,3)$ or $SL(2,5) \lhd G_0$, then $S = AG(2,q)$, $v = q^2$; if G_0 has a normal extraspecial subgroup E of order 2^5 and G_0/E is isomorphic to a subgroup of S_5 then $S = AG(4,3)$ or $AG(2,9)$, or S is the nearfield affine plane of order 9; if $SL(2,13) = G_0$, $v = 3^6$, then $S = AG(6,3)$, the Hering affine plane of order 27 or one of the two Hering Steiner systems $S(2,9,9^3)$. □

Theorem 8.4.2. (Delandtsheer, Doyen, Siemons and Tamburini 1986.) If S is a finite linear space with Aut(S) 2-homogeneous but not 2-transitive on points, then $v = q^n$, $q \equiv 3 \pmod 4$, n odd, and S is:

(a) an affine space over a subfield of $GF(q^n)$, or
(b) a Netto system, $k = 3$, $q \equiv 7 \pmod{12}$.

We do not give the proof of the above result, as it relies heavily on several purely group theoretic results.

Theorem 8.4.3. (Delandtsheer 1984a.) If S is an antiflag transitive finite linear space, then S is

(a) a Desarguesian affine or projective space,

(b) the nearfield affine plane of order 9, or the Hering affine plane of order 27 or

(c) a Hermitian unital.

Once again, the proof is omitted because of its heavy reliance on group theoretic results.

Theorem 8.4.4. (Delandtsheer 1986.) Let S be a finite linear space which contains properly some non-trivial linear subspace, and which has some line of size greater than 2. If Aut(S) is homogeneous on pairs of intersecting lines, then S is affine or projective.

PROOF. Let π be a minimal proper non-trivial linear subspace of S. Define the *planes* of S to be all images under $G = \text{Aut}(S)$ of π. By the homogeneity hypothesis, any two intersecting lines are in at least one plane. If L_1 and L_2 were intersecting lines each in the distinct planes $g_1(\pi)$ and $g_2(\pi)$, then $g_1^{-1}(g_1(\pi) \cap g_2(\pi))$ would be a proper non-trivial linear subspace of π, contradicting the minimality of π. Hence, L_1 and L_2 are in a unique plane. It follows that S provided with its planes forms a planar space.

For any point p of S, the local space S_p is a non-trivial linear space with line size ≥ 3, and the group G^p induced on S_p by the stabilizer G_p acts 2-homogeneously on the points of S_p. By Theorems 8.4.1 and 8.4.2 it follows that one of the following occurs. S_p is

(i) a Desarguesian affine space of dimension ≥ 2, or the nearfield affine plane of order 9, or the Hering affine plane of order 27, or one of the two Hering systems $S(2,9,9^3)$;

(ii) a Desarguesian projective space $PG(d,q)$ with $G^p \supseteq PSL(d+1,q)$, $d \geq 2$, or $G^p \approx A_7$ inside $PSL(4,2)$, or $G^p = \text{Frob}(21)$ inside $PSL(3,2)$;

(iii) a Hermitian unital of order q, $G^p \supseteq PSU(3,q)$;

(iv) a Ree unital of order q, $G^p \supseteq {}^2G_2(q)$;

(v) a Netto triple system.

It is not difficult to see that for each p, G_p is transitive on the lines on p, and consequently all lines of S have the same size, which we shall denote by k.

Proposition 7.4.1 states that if each S_p is projective, then S is a dimensional linear space and S_p is isomorphic to S_q for all points p and q of S. In fact, it is not difficult to generalize this result to the following here: If S_p is projective or affine, then S is a dimensional linear space and S_p is isomorphic to S_q for all points q of S.

Suppose then that for some point p, S_p is affine of dimension ≥ 2. P. J. Cameron (1974) proved that a finite planar space in which all lines have the same size k and such that each S_p is an affine plane is necessarily an inversive plane. But, as mentioned earlier, this is a linear space in which each line has precisely two points, contradicting one of our assumptions. Take a 3-space V on p. Then V_p is an affine plane, and it follows that $k = 2$.

Suppose that for some point p, S_p is projective. Then each S_p is projective and isomorphic to $PG(d,q)$, $d \geq 2$, $q \geq 2$. Theorems 7.4.2 and 7.4.3 imply that S is affine or projective, or $d = 2$ and $q = k^2$ or $k^3 + k$. Since q must be a prime power, $q = k^3 + k$ is not possible. Consider the case $q = k^2$. From (ii) above, $G^p \supseteq PSL(3,q)$, which is transitive on the triples of collinear points of $PG(2,q)$. Thus G is transitive on the triples of concurrent coplanar lines of S. Let L_1 and L_2 be lines of S intersecting in the point p. We count in two ways the pairs (L,L') where L is a line meeting L_1 and L_2 not in p, and L' is a line on p meeting L, $L' \neq L_1, L_2$. Letting t be the number of choices for L, we have $(q - 1)t = (k - 1)^2(k - 1)$. Since $q = k^2$, this implies $k + 1|6$. Since $k \geq 3$, it follows that $k = 5$, $q = 25$ and $t = 2$.

If for some triple of distinct concurrent coplanar lines, there is an element of G_p, p the point of intersection of these lines, which interchanges the two trisecants, then this is true for all such triples because of the transitivity of G. Now let L and L' be any two lines not on the point p. Let L_1, L_2, L_3 and L_1', L_2', L_3' be triples on p meeting L and L' respectively. Then there is a map of G_p taking L to a trisecant M of L_1', L_2', L_3'. If $M \neq L'$, there is a map of G_p taking M to L'. So in any case, we can map L to L' using G_p. Thus G_p has a unique orbit on the lines of S not through p.

Since G_p is also transitive on lines through p, G_p has exactly two line orbits on S. By Theorem 8.2.2, G_p has exactly two point orbits on S. Since G is transitive on points, it follows that G is 2-transitive on points. By Theorem 8.4.1, S is an affine or projective space, or a unital. But $|S| = 1 + (k - 1)(q^2 + q + 1) = 2605$, which leads us to a contradiction.

Thus G_p is not transitive on points of $S - p$, and no element of G_p interchanges the two trisecants of a triple of distinct coplanar lines on p.

Let O and O' be distinct orbits of G_p on $S - p$. Let L and L' be lines on p such that $L \cap O \ni q$ and $L' \cap O' \ni q'$. Let $M = qq'$, and let M' be a line on p meeting M in x, $x \neq q,q'$. There is an element $g \in G_p$

fixing M' and interchanging L and L'. Thus, since g cannot map M to the other trisecant of L, L' and M', g maps M to M and therefore interchanges q and q'. This contradicts the fact that O and O' are distinct orbits.

We have to this point completely covered cases (i) and (ii) above.

Suppose that for some p, S_p is a Hermitian or Ree unital. Then S_p has $q^3 + 1$ points, $k = q + 1$ and q is a prime power. Let v and v_π be respectively the numbers of points of S and of the plane π of S. Then

$$v_\pi - 1 = (q + 1)(k - 1)$$
$$v - 1 = (q^3 + 1)(k - 1) = (q^2 - q + 1)(v_\pi - 1).$$

Since each plane of S contains exactly $v_\pi(v_\pi - 1)/k(k - 1)$ lines, we have $k(k - 1)|v_\pi(v_\pi - 1)$ and therefore $k|((q + 1)(k - 1) + 1)(q + 1)$, giving $k|q(q + 1)$.

The number of planes on p is $q^2(q^2 - q + 1)$, and so the total number of planes in S is the integer $vq^2(q^2 - q + 1)/v_\pi$. Therefore

$$(q + 1)(k - 1) + 1 \,|q^3(q - 1)(q^2 - q + 1). \tag{1}$$

If S_p is a Hermitian unital, then $G^p \supseteq PSU(3,q)$, which is transitive on the triples of collinear points of S_p. Arguing as above, we obtain the equation

$$(k - 1)^2 (k - 2) = (q - 1)t$$

where t is the constant number of trisecants of three concurrent coplanar lines.

If S_p is a Ree unital, then $G^p \supseteq \Gamma \approx {}^2G_2(q)$. Recall that the stabilizer under Γ of two points has order $q - 1$ and of three collinear points has order 2. Therefore the stabilizer of two points of S_p in Γ has two orbits of size $(q - 1)/2$ on the line containing these two points. Hence, given three coplanar lines all intersecting in p there are either t_1 or t_2 trisecants of the three lines for fixed integers t_1 and t_2. It follows that $(k - 1)^2(k - 2) = (t_1 + t_2)(q - 1)/2$. From this equation and the one above for the Hermitian unital, we conclude that in any case,

$$q - 1|2(k - 1)^2 (k - 2). \tag{2}$$

Let $m = (q + 1)(k - 1) + 1 = q(k - 1) + k$. Then

$$(m,q - 1) = ((q - 1)(k - 1) + 2k - 1, q - 1)|2k - 1.$$

On the other hand, condition (2) implies

$$(m,q - 1)|(m,2(k - 1)^2(k - 2))|(m,2(k - 2))|2k - 4.$$

And therefore $(m,q - 1)|3$. From condition (1) we now obtain

$$m|3q^3(q^2 - q + 1) = 3q^3[(q - 2)m - q + 3k - 1],$$

and therefore

$$m|3q^3(q + 1 - 3k). \tag{3}$$

Clearly, $k \leq q + 1$. If $k = q + 1$, then all planes are projective, and so S is projective (see Theorem 7.3.2). If $k = q > 3$, then S is an affine space by Theorem 7.3.1. In either case, we contradict the fact that S_p is a unital. If $k = q = 3$, then S is a Hall triple system (Section 7.3), and J. I. Hall (1972) shows that v is a power of 3, which is impossible. Therefore $k < q$.

Suppose $(k,q) = 1$. Then since $k|q(q + 1)$, we have $(m,q) = 1$ also. Thus by condition (3), $mk|3(q + 1 - 3k)$. If $q + 1 - 3k > 0$, then $3m < km + 9k \leq 3(q + 1)$, and so $k(q + 1) < 2q + 1$, contradicting $3 \leq k$. If $q + 1 - 3k < 0$, we have $(m - 9)k + 3q + 3 \leq 0$, implying $q(k - 1) < 9 - k$, and contradicting $3 \leq k < q$. If $q + 1 - 3k = 0$, then from condition (2), $3k - 2|2(k - 1)^2(k - 2)$, while $(3k - 2, k - 1) = 1$, $(3k - 2, k - 2)|8$ imply $3k - 2|8$, contradicting $k \geq 3$.

Thus $(k,q) \neq 1$, and we may write $q = c^n$, $k = c^\alpha h$, c a prime, h, α and n integers, $c|h$, $1 \leq \alpha \leq n$, and $1 \leq h < c^{n-\alpha}$.

Now using condition (3) we obtain

$$(c^n + 1)h - c^{n-\alpha}|3(c^{n-\alpha} - 3c^\alpha h^2). \tag{4}$$

First of all we eliminate some small values for $c^{n-\alpha}$.

Suppose that $c^{n-\alpha} = 2$. Then $c = 2$, $h = 1$ and $k = 2^{n-1}$. From equation (2), $2^n - 1|2(2^{n-1} - 1)^2(2^{n-1} - 2)|(2^n - 2)^2(2^n - 4)$, implying $2k - 1 = 2^n - 1|3$, a contradiction.

Suppose that $c^{n-\alpha} = 4$. Then $c = 2$, $h = 1$ or 3 and $k = 2^{n-2}$ or $3 \cdot 2^{n-2}$. From (2), $2^n - 1|2(2^{n-2}h - 1)^2(2^{n-2}h - 2)|(2^nh - 4)^2(2^nh - 8)$, implying $2^n - 1|(h - 4)^2(h - 8)$. The only possibility is $q = 64$, $k = 16$, which is impossible by condition (3).

Suppose that $c^{n-\alpha} = 5$. Then $c = 5$, $h = 1,2,3$ or 4 and $k = 5^{n-1}h$. From condition (2), $5^n - 1|2(5^{n-1}h - 1)^2(5^{n-1}h - 2)|2(5^nh - 5)^2(5^nh - 10)$, implying $5^n - 1|2(h - 5)^2(h - 10)$. So $5^n - 1 \leq 288$. The only possibilities are $5^n = 25$ or 125. Thus $q = 125$ and $k = 5$ or 10. Both of these contradict condition (3).

Suppose that $q > 3k^2$. or equivalently, $c^{n-\alpha} > 3c^\alpha h^2$. Since $\alpha \geq 1$ and $(c,h) = 1$, condition (4) yields $(c^n + 1)h - c^{n-\alpha}|3(c^{n-\alpha-1} - 3c^{\alpha-1}h^2)$. Hence $c^nh < c^nh + h + 9c^{\alpha-1}h^2 \leq (3 + c)c^{n-\alpha-1}$, and so $c(c^\alpha h - 1) < 3$, which implies $k = 2$, and a contradiction.

Suppose that $q < 3k^2$, or equivalently, $c^{n-\alpha} < 3c^\alpha h^2$. Set $\beta =$

$\min\{\alpha, n-\alpha\} \geq 1$. By condition (4), $c^\beta(c^n + 1)h - c^{n+\beta-\alpha}|9c^\alpha h^2 - 3c^{n-\alpha}$. So $c^\beta(c^n + 1)h \leq 9c^\alpha h^2 + (c^\beta - 3)c^{n-\alpha}$, which is $<3c^\beta c^\alpha h^2$ if $c^\beta > 3$.

If $c^\beta > 3$, then $c^n + 1 < 3c^\alpha h$. But if $c^\beta = 3$, we get $c = 3$ and $3[(3^n + 1)h - 3^{n-\alpha}]|9(3^\alpha h^2 - 3^{n-\alpha-1})$. Therefore $(3^n + 1)h - 3^{n-\alpha}|3^\alpha h^2 - 3^{n-\alpha-1}$, and so $(3^n + 1)h \leq 3^\alpha h^2 + 2\cdot 3^{n-\alpha-1} < 3\cdot 3^\alpha h^2$. Thus we also obtain $c^n + 1 < 3c^\alpha h$.

Therefore, suppose $c^n + 1 < 3c^\alpha h$ and $c^\beta \neq 2$. Recalling that $(c^n + 1)h - c^{n-\alpha}|3(c^n + 1 - 3c^\alpha h)$, we have $(c^n + 1)h - c^{n-\alpha} \leq 9c^\alpha h - 3(c^n + 1)$, and so, since $q + 1 > k$, we have $(c^n - 2)h < (c^n - 2)h + (3h - c^{n-\alpha}) \leq 9c^\alpha h - 3(c^n + 1) < 6c^\alpha h$. So $c^\alpha(c^{n-\alpha} - 6) < 2$. But $c^\alpha \geq 2$, and we eliminated earlier the possibilities $c^{n-\alpha} = 2, 4, 5$. Thus $c^{n-\alpha} = 3$. But as for the case $c^\beta = 3$ above, this implies $(3^n + 1)h \leq 3^{n-1}h^2 + 2 = kh + 2$. Since $(k + 2)h \leq (q + 1)h = (3^n + 1)h$, we conclude that $h = 1$ and $3^n + 1 = k + 2$, contradicting $3|k$.

We now turn to the case $c^\beta = 2$. Since $c^{n-\alpha} \neq 2, 4, 5$, it follows that $c^\alpha = 2$. Thus $(m,q) = (k,q) = 2$. Therefore $m \equiv k \equiv 2 \pmod 4$, $k \geq 6$, and condition (3) becomes $m|6(q + 1 - 3k)$. Both q and k are even, so that $q + 1 - 3k \neq 0$. If $q + 1 > 3k$, we obtain $qk - q + k \leq 6(q + 1 - 3k)$, contradicting $k \geq 6$. If $q + 1 < 3k$, then $qk - q + k \leq 6(3k - q - 1)$, so that $5q + 6 + k(q - 17) \leq 0$. It follows that $q < 17$. The only possibilities since $q > k \geq 6$, are $q = 8, 16$, which do not satisfy $m|6(q + 1 - 3k)$.

Finally we consider $q = 3k^2$. Then $k = 3^\alpha$, $q = 3^{2\alpha+1}$, and condition (2) becomes $3^{2\alpha+1} - 1|2(3^\alpha - 1)^2(3^\alpha - 2) = 2(3^{3\alpha} - 4\cdot 3^{2\alpha} + 5\cdot 3^\alpha - 2)$, implying $3^{2\alpha+1} - 1|2(3^{3\alpha} - 3^{2\alpha} + 5\cdot 3^\alpha - 3)$, in turn, implying $3^{2\alpha+1} - 1|2(-3^{2\alpha} + 16\cdot 3^{\alpha-1} - 3)$, and then $3^{2\alpha+1} - 1|2(16\cdot 3^\alpha - 10)$, from whence $3k^2 - 32k + 19 \leq 0$. Thus $k = 3$ and $q = 27$, or $k = 9$ and $q = 243$. But neither of these satisfies condition (2).

We consider now the case where S_p is a Steiner system $S(2,9,9^3)$.

Let v be the number of points in S, v' be the number in any plane of S and $k = q + 1$ be the constant number of points per line in S. Then $v' - 1 = 9q$ and $v - 1 = 9^3q$. Since every point of S is on $9 > (9^3 - 1)/9(9 - 1) = 9^2\cdot 91$ planes, we have $v'|v\cdot 9^2\cdot 91$, or $9q + 1|(9^3q + 1)\cdot 9^2\cdot 91$.

Since in every plane of S, the points have degree 9, we have $3 \leq k = q + 1 \leq 9$. Thus $q = 3$. Therefore the planes of S are unitals $S(2,4,28)$.

The automorphism group of a plane π of S acts transitively on the unordered pairs of intersecting lines in π, and thus its order is divisible by 1008. This implies, by a result of P. J. Cameron (1974), that the planes are Hermitian unitals whose full automorphism group is $P\Gamma U(3,3)$.

Since all maximal subgroups of $PSU(3,3)$ have order ≤ 216 (see J. Fisher and J. McKay 1978), any subgroup of $P\Gamma U(3,3)$ whose order is divisible by 1008 contains $PSU(3,3)$. Hence for any plane π of S, the stabilizer G_π acts 2-transitively on the points of π, and so G is 2-transitive on the points of S. This contradicts Theorem 8.4.1.

We come finally to the case where S_p for each point p is a Netto triple system. In any plane of S, each point is on 3 lines, and so $k = 3$. Hence all planes of S are Fano planes, $S = PG(d,2)$ and $S_p = PG(d-1,2)$. Recall that the only Netto triple system which is also a projective space is $PG(2,2)$. Thus we have handled this case above. □

What happens if the 'dimension' requirement is dropped in Theorem 8.4.4? This is not known at present, and we leave it as a research problem.

In the next theorem, we do drop the dimensionality condition, but replace it by a second transitivity condition.

Theorem 8.4.5. (Delandtsheer 1984b.) Let S be a finite linear space whose automorphism group is (i) transitive on pairs of intersecting lines, and (ii) transitive on pairs of non-intersecting lines. Then S is a Desarguesian affine plane, a Desarguesian projective space or a complete graph.

PROOF. By Theorem 8.3.3, S is 2-transitive on points. Thus we may invoke Theorem 8.4.1. Affine spaces of dimension > 2 do not satisfy condition (ii). However, all other linear spaces appearing under part (a) of Theorem 8.4.1 do satisfy both conditions (i) and (ii).

If S is a Hermitian unital, and g is a Hermitian polarity of the projective plane $\pi = PG(2,q^2)$, there are pairs of lines of π (L,L') with $g(L) \in L'$, and (M,M') with $g(M) \notin M'$, which induce in the unital pairs of disjoint lines. Since every automorphism of S is induced by an automorphism of π, condition (ii) is not satisfied. (For more details, see M. E. O'Nan 1972 and D. E. Taylor 1974.)

If S is a Ree unital, it follows from Theorem 8.1.1 that for $q \neq 3$, $|\mathrm{Aut}({}^2G_2(q)')| = (2m+1)\cdot|{}^2G_2(q)'|$ and for $q = 3$, $G = P\Gamma L(2,8)$. For any line L, the number of lines disjoint from L is $q(q-1)(q^2-q-1)$. So using condition (ii), $q \neq 3$ gives $q(q-1)(q^2-q-1)|(2m+1)|G_L| = (2m+1)q(q-1)(q+1)$, a contradiction, since $q^2-q-1 > (2m+1)(q+1)$. But $q = 3$ implies $3\cdot 2\cdot 5|3\cdot 8\cdot 63$, also a contradiction.

Suppose S is the Hering affine plane, which was described at the beginning of this section. Using results in W. M. Kantor (1985), the stabilizer in Aut (A_{27}) of the point (0,0,0,0,0,0) and the line L generated by the vectors (1,0,0,0,0,0), (0,1,0,0,0,0) and (0,0,1,0,0,0) has order $2 \cdot 3 \cdot 7$, which is not divisible by 27, the number of lines distinct from L and on the above point. Thus condition (i) is not satisfied.

If S is the nearfield affine plane of order 9, J. André (1955) proved that the automorphism group G of S is not 2-transitive on the points of the line at infinity. Thus G_p, the stabilizer of a point p, is not 2-transitive on the lines through p, so that condition (i) does not hold.

If S is either of the two Steiner systems due to C. Hering, then the stabilizer of a point-line incident pair on the full automorphism group of these spaces has order $|SL(2,13)|/91 = 3 \cdot 2^3$, which is not divisible by 90. This implies that condition (i) is not satisfied. □

The following result is worth mentioning.

Theorem 8.4.6. (Delandtsheer 1984b). Let S be a finite linear space whose automorphism group is 2-transitive on lines. Then S is a Desarguesian projective plane or a triangle.

PROOF. It follows immediately that all lines of S intersect and all lines have the same size. So S is a projective plane or a triangle.

If S is a projective plane, the dual is a projective plane which is 2-transitive on points. By Theorem 8.1.2, S and its dual are Desarguesian. □

Our final result is the following impressive theorem.

Theorem 8.4.7. (Buekenhout, Delandtsheer, Doyen, Kleidman, Liebeck and Saxl 1990.) If $G \not\subseteq A\Gamma L(1, q)$ on S, is a flag transitive collineation group of an $S(2, k, v)$, $3 \leq k < v$, then either

1. *S has the structure of an affine space AG (d, p); G has a (unique) elementary abelian minimal normal subgroup T (the translation group of AG (d, p)) which acts regularly on the points of S, $|T| = p^d$, p a prime, $d \geq 1$, or*
2. *G has a nonabelian simple subgroup N such that $N \lhd G \subseteq \mathrm{Aut}\, N$ and*
   (i) $S = PG(d, q)$, $d \geq 2$ and $N = PSL(d + 1, q)$, or $S = PG(3, 2)$ and $N = A_7$,
   (ii) $S = U_H(q)$ is a Hermitian unital and $N = PSU(3, q)$,
   (iii) $S = U_R(3^{2e+1})$ is a Ree unital, $e \geq 1$, $N = {}^2G_2(3^{2e+1})$,

(iv) $S = W(2^e)$ is a Witt space, $e \geq 3$ and $N = PSL(2, 2^e)$.

In each case, the action of N on S is the usual one.

8.5 Exercises

1. Compute the automorphism group of the projective plane of order 3, and determine the point and line orbits. Find a linear space with $v_G > b_G + 1$.
2. In Example 8.2.1, prove in detail that there is a unique line orbit.
3. Show that the following statement, similar to that of Theorem 8.2.4, is *not* true: For all integer pairs (v,b) with $v \leq b$, there is a finite linear space with v points and b lines.
4. Classify those finite linear spaces which are (a) transitive on triples (p,p',p'') of noncollinear points; (b) transitive on triples (p,p',L), $p \notin L$, $p' \in L$.

8.6 Research problems

1. For which values of v_G does there exist a projective plane with $v_G = b_G$?
2. Classify those finite linear spaces for which $v_G = b_G$.
3. Does there exist an infinite linear space for which $v_G > 2$ and $b_G = 1$?
4. (Due to F. DeClerck.) Find those linear spaces for which $G \subset \text{Aut}(S)$ implies $v_G < v_{\text{Aut}(S)}$ (respectively, $b_G < b_{\text{Aut}(S)}$).
5. Let S be a finite linear space with Aut(S) homogeneous on pairs of intersecting lines. Describe S. (See Theorem 8.4.4.)

8.7 References

André, J. (1955), Projektive Ebenen über Fastkörpern. *Math. Z. 62* 137–160.

Batten, L. M. (1986), *Combinatorics of Finite Geometries*, Cambridge University Press, Cambridge, New York.

Block, R. E. (1967), On the orbits of collineation groups. *Math Z. 96* 33–49.

Blokhuis, A., Brouwer, A., Delandtsheer, A. and Doyen, J. (1987) Orbits on points and lines in finite linear and quasilinear spaces. *J. Comb. Theory* (A) 159–163.

Buekenhout, F. (1968), Homogénéité des espaces linéaires et des systèmes de blocs. *Math. Z. 104* 144–146.

Buekenhout, F., Delandtsheer, A. and Doyen, J. (1988), Finite linear spaces with flag-transitive groups. *J. Comb. Theory* (A) *49* 268–293.

Buekenhout, F., Delandtsheer, A., Doyen, J., Kleidman, P. B., Liebeck, M.

W. and Saxl, J. (1990), Linear spaces with flag-transitive automorphism groups. *Geom. Ded. 36* 89–94.

Cameron, P. J. (1974), Locally symmetric designs. *Geom. Ded. 3* 65–76.

Camina, A. R. and Gagen, T. M. (1984), Block transitive automorphism groups of designs. *J. of Algebra 86* 549–554.

Delandtsheer, A. (1984a) Finite antiflag-transitive linear spaces. *Mitt. Math. Sem.* Giessen *164* 65–75; erratum *166* (1984) 205.

Delandtsheer, A. (1984b) Transitivity on ordered pairs of lines in finite linear spaces. *Abh. Math. Sem. Univ. Hamburg 54* 107–110.

Delandtsheer, A. (1986), A geometric consequence of the classification of finite doubly transitive groups. *Geom. Ded. 21* 145–156.

Delandtsheer, A., Doyen, J., Siemons, J. and Tamburini, C. (1986), Doubly homogeneous 2-(v,k,1) designs. *J. Comb. Theory* (A) *43* 140–145.

Dembowski, P. (1968), *Finite Geometries*, Springer-Verlag Berlin, Heidelberg, New York.

Fisher, J. and McKay, J. (1978), The nonabelian simple groups G, $|G| < 10^6$-maximal subgroups. *Math. Comp. 32* 1293–1302.

Gorenstein, D. (1982), *Finite simple groups: An introduction to their classification*. Plenum, New York.

Hall, J. I. (1972), *Steiner Triple Systems and 2-Transitive Groups*. M. Sc. Diss. Oxford University.

Hering, C. (1969), Eine nicht-desarguessche zweifach transitive affine Ebene der Ordnung 27. *Abh. Math. Sem. Univ. Hamburg 34* 203–208.

Hering, C. (1985), Two new sporadic doubly transitive linear spaces. *Finite Geometries Lecture Notes in Pure and Applied Mathematics 103*. Marcel Dekker, New York 127–129.

Kantor, W. M. (1972), On 2-transitive groups in which the stabilizer of two points fixes additional points. *J. London Math. Soc.* (2) *5* 114–122.

Kantor, W. M. (1985), Homogeneous designs and geometric lattices. *J. Comb. Theory* (A) *38* 66–74.

Lüneburg, H. (1966), Some remarks concerning the Ree groups of type (G_2). *J. of Algebra 3*, 256–259.

O'Nan, M. E. (1972), Automorphisms of unitary block designs. *J. of Algebra 20* 495–511.

Ostrom, T. G. (1958), Dual transitivity in finite projective planes. *Proc. A. M. S. 9* 55–56; correction *Canad. J. Math. 10* 507–512.

Ostrom, T. G. and Wagner, A. (1959), On projective and affine planes with transitive collineation groups. *Math. Z. 71* 186–199.

Ree, R. (1961a), A family of simple groups associated with the simple Lie algebra of type (F_4). *Amer. J. Math. 83* 401–420.

Ree, R. (1961b), A family of simple groups associated with the simple Lie algebra of type (G_2). *Amer. J. Math. 83* 432–462.

Robinson, R. M. (1975), The structure of certain triple systems. *Math. Comp. 29* 223–241.

Saxl, J. (1981), On points and triples of Steiner triple systems. *Arch. Math. 36* 558–564.

Saxl, J. (1983), Groups on points and sets. *Ann. Discr. Math. 18* 721–724.

Taylor, D. E. (1974), Unitary block designs. *J. Comb. Theory* (A) *16* 51–56.

Tits, J. (1960), Les groupes simples de Suzuki et de Ree. *Séminaire Bourbaki* 3e année. No 210.

Appendix

The following pages contain descriptions of the linear spaces on at most nine points. For each linear space, we give a picture, the number of lines, the collineation group acting on the space (this appears in a box) and the number of point orbits under this group.

We use the following group theoretic notations for the collineation groups:

C_n is the cyclic group of order n,
S_n is the symmetric group on n letters,
D_{2n} is the dihedral group of order $2n$,
$\Pi_i H$ is the direct product of i groups H,
$\Sigma_i H$ is the direct sum of i groups H,
G_n is a group of order n.

The general notation G is used when the group is not describable as a direct sum or product of the cyclic, symmetric or dihedral groups.

We wish to thank Jean Doyen, who, with much time and effort, compiled this comprehensive table for us and offered to let us use it as an appendix to this monograph. Any errors in it are due to the present authors.

0 point

Graph	b	Group	
ϕ	$b = 0$		

1 point

Graph	b	Group	
	$b = 0$	G_1	1

2 points

Graph	b	Group	
	$b = 1$	C_2	1

3 points

Graph	b	Group	
	$b = 1$	S_3	1
	$b = 3$	S_3	1

4 points

Graph	b	Group	
	$b = 1$	S_4	1
	$b = 4$	S_3	2
	$b = 6$	S_4	1

5 points

Graph	b	Group	
	$b = 1$	S_5	1
	$b = 5$	S_4	2
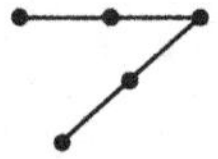	$b = 6$	D_8	2
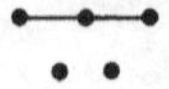	$b = 8$	$S_3 \times C_2$	2
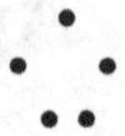	$b = 10$	S_5	1

6 points

Graph	b	Group	
	$b = 1$	S_6	1
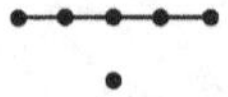	$b = 6$	S_5	2
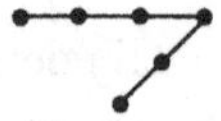	$b = 8$	$S_3 \times C_2$	3
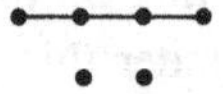 	$b = 10$	$S_4 \times C_2$	2
	$b = 7$	S_4	1
	$b = 9$	S_3	2
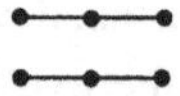	$b = 11$	$\sum_2 S_3$	1
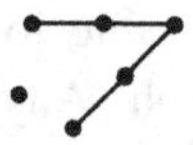	$b = 11$	D_8	3
	$b = 13$	$\prod_2 S_3$	2
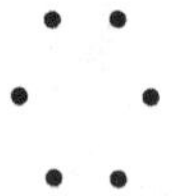	$b = 15$	S_6	1

7 points

b	Group	
 $b = 1$	S_7	1
 $b = 7$	S_6	2
 $b = 10$	$S_4 \times C_2$	3
 $b = 12$	$S_5 \times C_2$	2
 $b = 11$	$\sum_2 S_3$	2
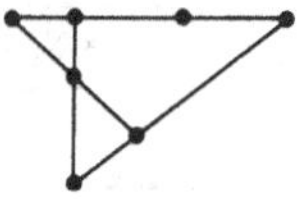 $b = 10$	S_3	3
 $b = 12$	$C_2 \times C_2$	4
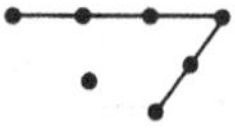 $b = 14$	$S_3 \times C_2$	4
 $b = 14$	$S_4 \times S_3$	2
 $b = 16$	$S_4 \times S_3$	2
 $b = 7$	G_{168}	1
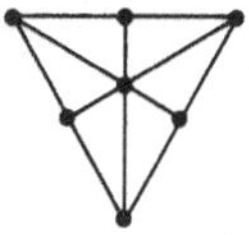 $b = 9$	S_4	2
 $b = 11$	D_8	3
 $b = 11$	$S_3 \times C_2$	2
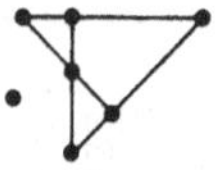 $b = 13$	S_4	2
 $b = 13$	$C_2 \times C_2$	3
 $b = 13$	S_3	3
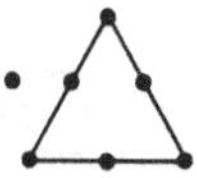 $b = 15$	S_3	3
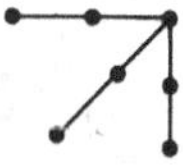 $b = 15$	$\sum_3 C_2$	2
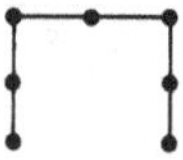 $b = 15$	D_8	3
 $b = 17$	$\sum_2 S_3$	2
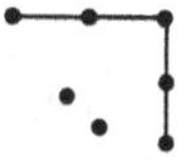 $b = 17$	$D_8 \times C_2$	3
 $b = 19$	$S_4 \times S_3$	2
 $b = 21$	S_7	1

8 points

b	Group	
 $b = 1$	S_8	1
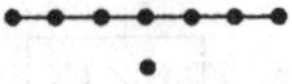 $b = 8$	S_7	2
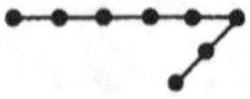 $b = 12$	$S_5 \times C_2$	3
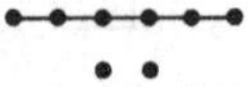 $b = 14$	$S_6 \times C_2$	2
 $b = 14$	$S_4 \times S_3$	3
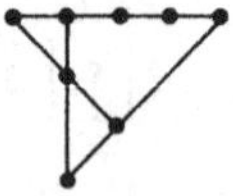 $b = 13$	$S_3 \times C_2$	3
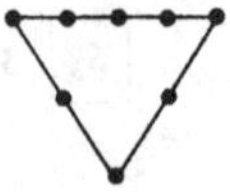 $b = 15$	$S_3 \times C_2$	4
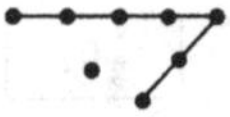 $b = 17$	$S_4 \times C_2$	4
 $b = 17$	$S_5 \times S_3$	2
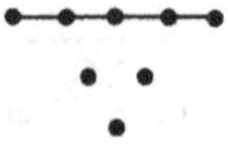 $b = 19$	$S_5 \times S_3$	2
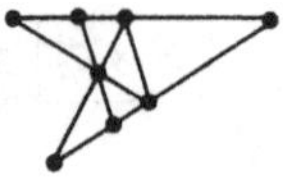 $b = 12$	$S_3 \times C_2$	3
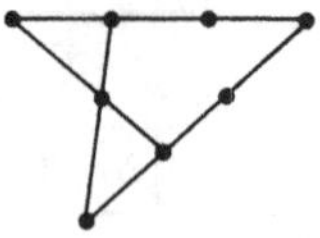 $b = 14$	$C_2 \times C_2$	4
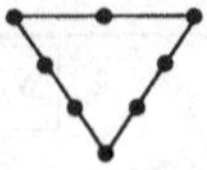 $b = 16$	D_8	4
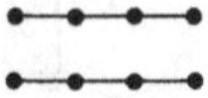 $b = 18$	$\sum_2 S_4$	1
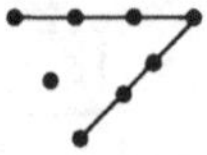 $b = 18$	$\sum_2 S_3$	3
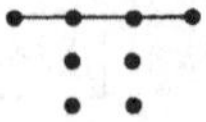 $b = 23$	$S_4 \times S_4$	2
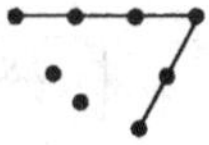 $b = 21$	$S_3 \times C_2 \times C_2$	4
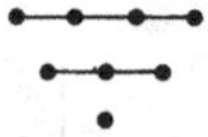 $b = 21$	$S_4 \times S_3$	3
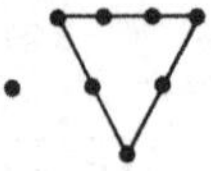 $b = 19$	$C_2 \times C_2$	5
 $b = 19$	$S_3 \times D_8$	3
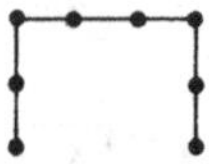 $b = 19$	$D_8 \times C_2$	3
 $b = 19$	$S_3 \times C_2$	5
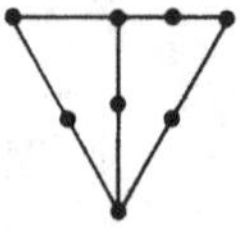 $b = 17$	S_3	4
 $b = 17$	$C_2 \times C_2$	5

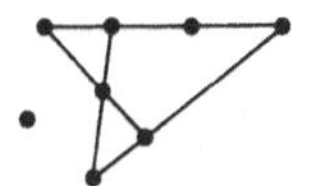

$b = 17$ $\boxed{S_3}$ 4

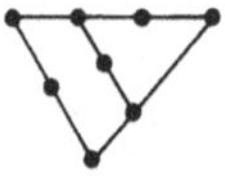

$b = 17$ $\boxed{C_2}$ 5

$b = 17$ $\boxed{C_2 \times C_2}$ 5

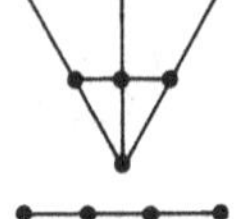

$b = 15$ $\boxed{S_3}$ 4

$b = 15$ $\boxed{C_2}$ 6

$b = 15$ $\boxed{C_2 \times C_2}$ 4

$b = 15$ $\boxed{C_2}$ 6

$b = 15$ $\boxed{D_8 \times C_2}$ 3

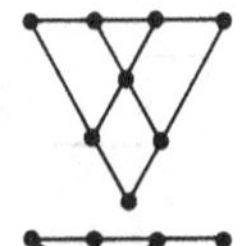

$b = 15$ $\boxed{D_8}$ 2

$b = 13$ $\boxed{C_2}$ 5

$b = 13$ $\boxed{C_2 \times C_2}$ 5

$b = 11$ $\boxed{S_4}$ 3

$b = 11$ $\boxed{D_8}$ 3

$b = 28$ $\boxed{S_8}$ 1

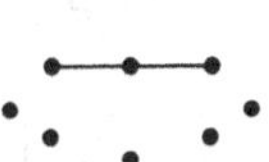

$b = 26$ $\boxed{S_3 \times S_5}$ 2

$b = 24$ $\boxed{C_2 \times \sum_2 S_3}$ 2

$b = 24$ $\boxed{S_3 \times D_8}$ 3

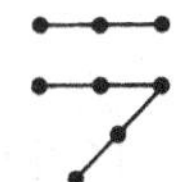

$b = 22$ $\boxed{S_3 \times D_8}$ 3

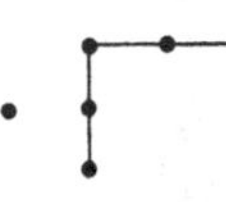

$b = 22$ $\boxed{D_8}$ 4

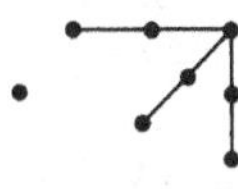

$b = 22$ $\boxed{\sum_3 C_2}$ 3

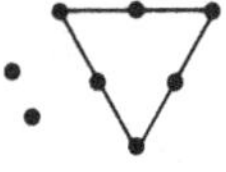

$b = 22$ $\boxed{S_3 \times C_2}$ 3

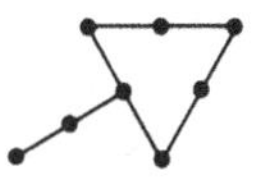

$b = 20$ $\boxed{C_2 \times C_2}$ 5

b	Group	
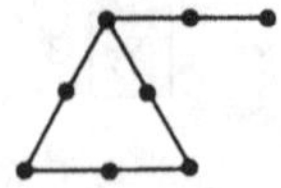 $b = 20$	$C_2 \times C_2$	5
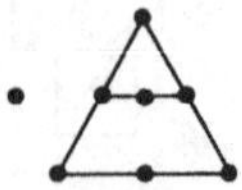 $b = 20$	$C_2 \times C_2$	4
 $b = 20$	D_8	2
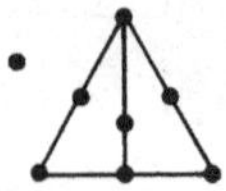 $b = 20$	S_3	4
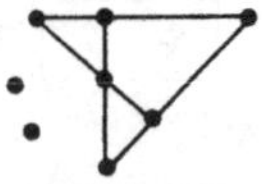 $b = 20$	$S_4 \times C_2$	2
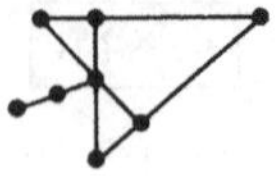 $b = 18$	$\prod_3 C_2$	4
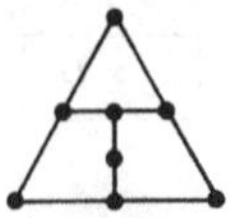 $b = 18$	$C_2 \times C_2$	4
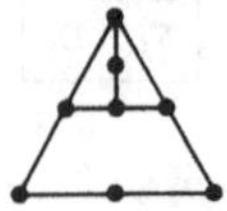 $b = 18$	C_2	6
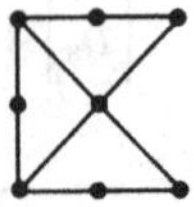 $b = 18$	C_2	5
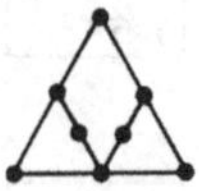 $b = 18$	C_2	5
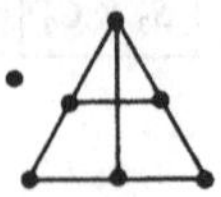 $b = 18$	$S_3 \times C_2$	3
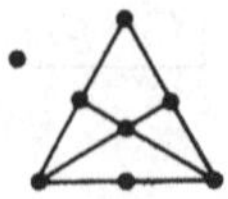 $b = 18$	D_8	4
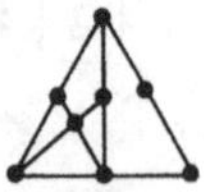 $b = 16$	C_2	6
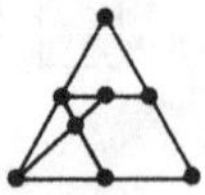 $b = 16$	C_2	5
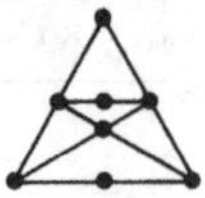 $b = 16$	D_8	3
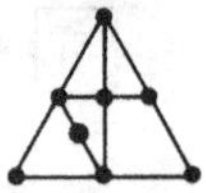 $b = 16$	C_2	5
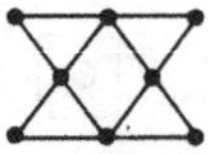 $b = 16$	$S_3 \times C_2$	2
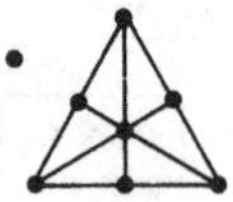 $b = 16$	S_4	3
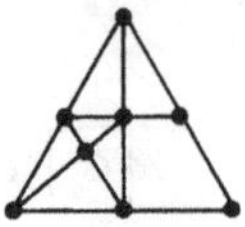 $b = 14$	$C_2 \times C_2$	4
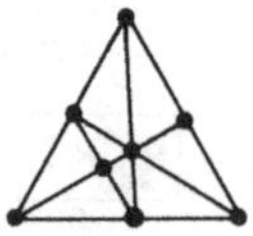 $b = 14$	D_8	4
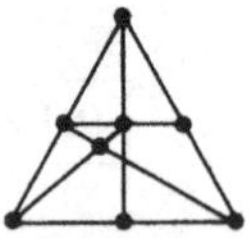 $b = 14$	$C_2 \times C_2$	3
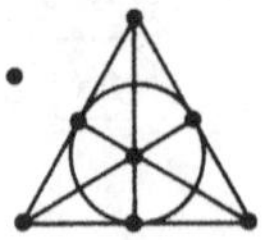 $b = 14$	G_{168}	2
 $b = 12$	G_{48}	1

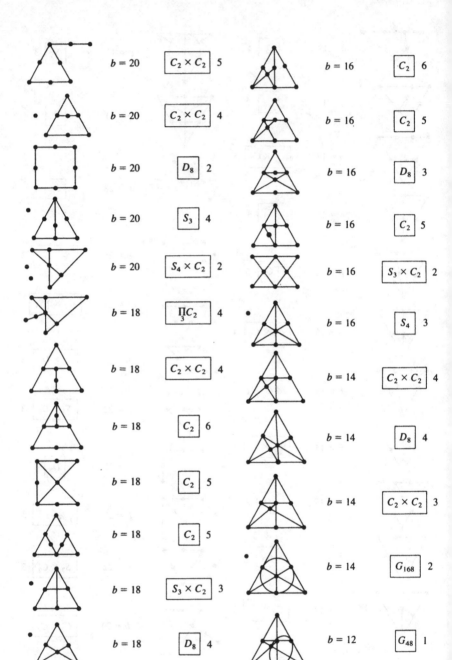

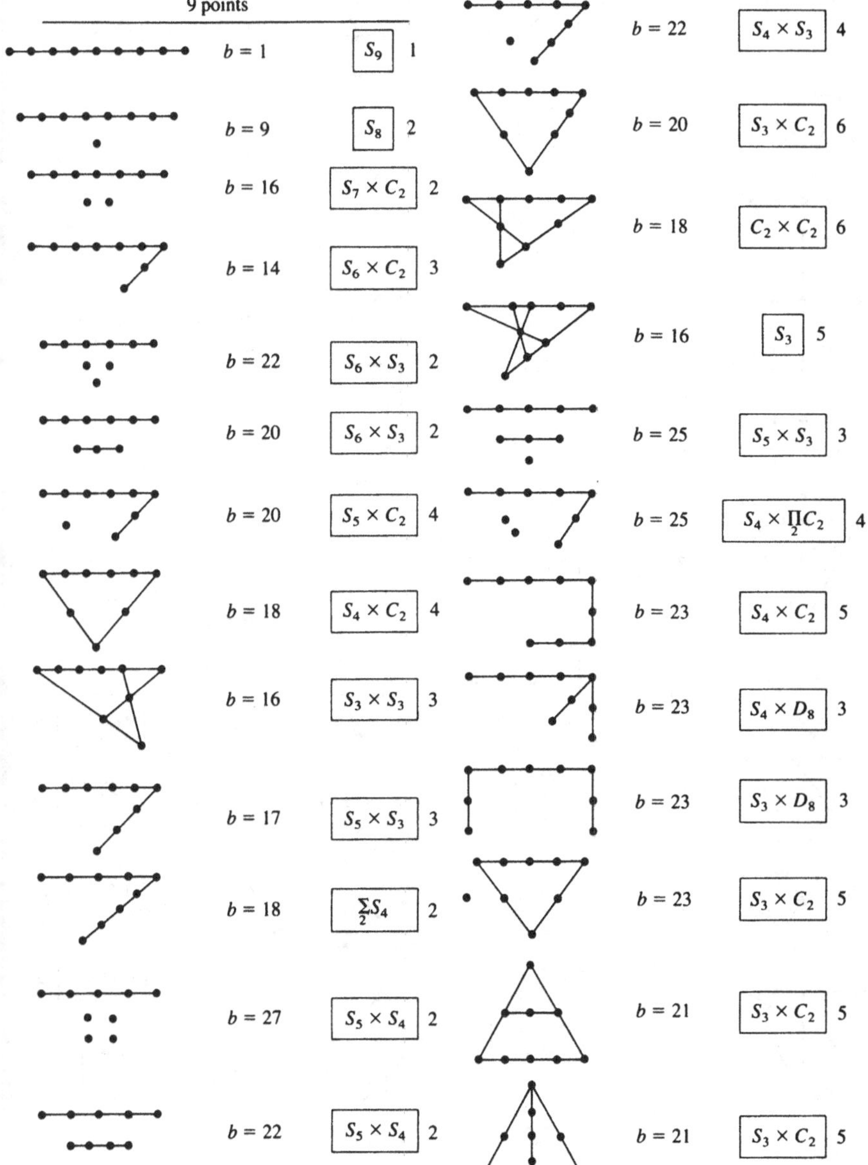
9 points
b = 1
S9
1
b = 9
S8
2
b = 16
S7 × C2
2
b = 14
S6 × C2
3
b = 22
S6 × S3
2
b = 20
S6 × S3
2
b = 20
S5 × C2
4
b = 18
S4 × C2
4
b = 16
S3 × S3
3
b = 17
S5 × S3
3
b = 18
∑2 S4
2
b = 27
S5 × S4
2
b = 22
S5 × S4
2
b = 22
S4 × S3
4
b = 20
S3 × C2
6
b = 18
C2 × C2
6
b = 16
S3
5
b = 25
S5 × S3
3
b = 25
S4 × ∏2 C2
4
b = 23
S4 × C2
5
b = 23
S4 × D8
3
b = 23
S3 × D8
3
b = 23
S3 × C2
5
b = 21
S3 × C2
5
b = 21
S3 × C2
5

Graph	b	Group	
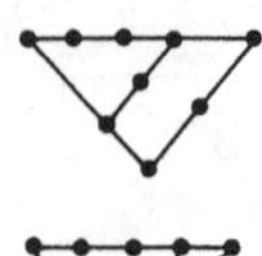	$b = 21$	$C_2 \times C_2$	5
	$b = 21$	$S_3 \times C_2$	4
	$b = 21$	$S_3 \times C_2$	4
	$b = 19$	$S_3 \times C_2$	4
	$b = 19$	$C_2 \times C_2$	6
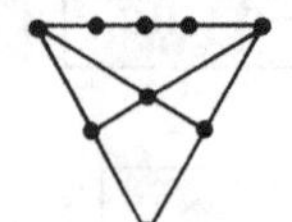	$b = 19$	$D_8 \times S_3$	3
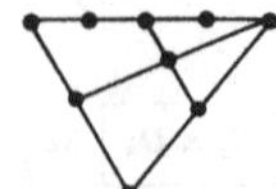	$b = 19$	$\prod_3 C_2$	4
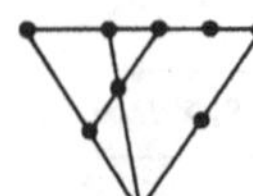 	$b = 19$	C_2	7
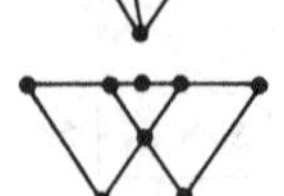 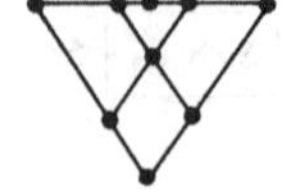	$b = 19$	D_8	3
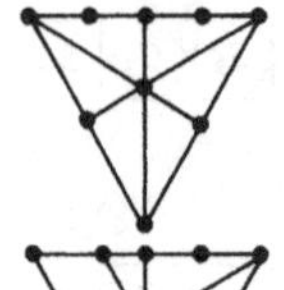	$b = 17$	$\prod_3 C_2$	5
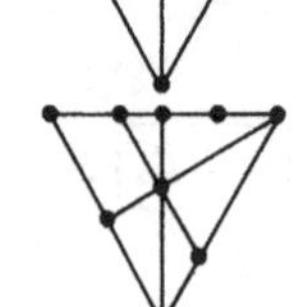	$b = 17$	C_2	6
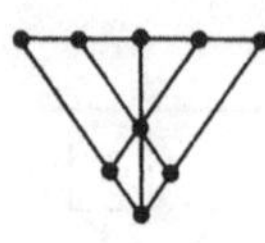	$b = 17$	$C_2 \times C_2$	4
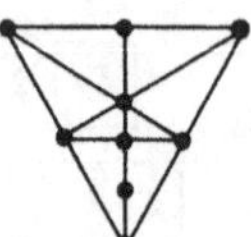	$b = 15$	D_8	4
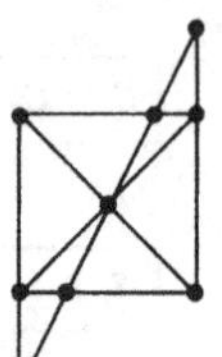	$b = 15$	D_8	3
 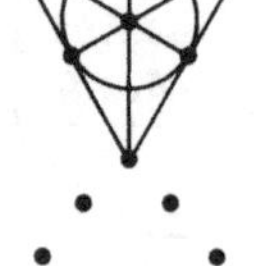	$b = 15$	$S_4 \times C_2$	3
	$b = 36$	S_9	1
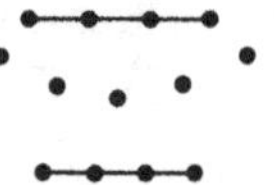	$b = 31$	$S_4 \times S_5$	2
	$b = 26$	$\sum_2 S_4$	2
	$b = 26$	$C_2 \times \sum_2 S_3$	3
	$b = 21$	$\sum_3 C_2$	2
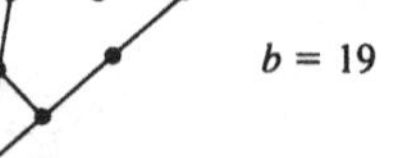	$b = 19$	S_3	3
	$b = 17$	$S_3 \times C_2$	2

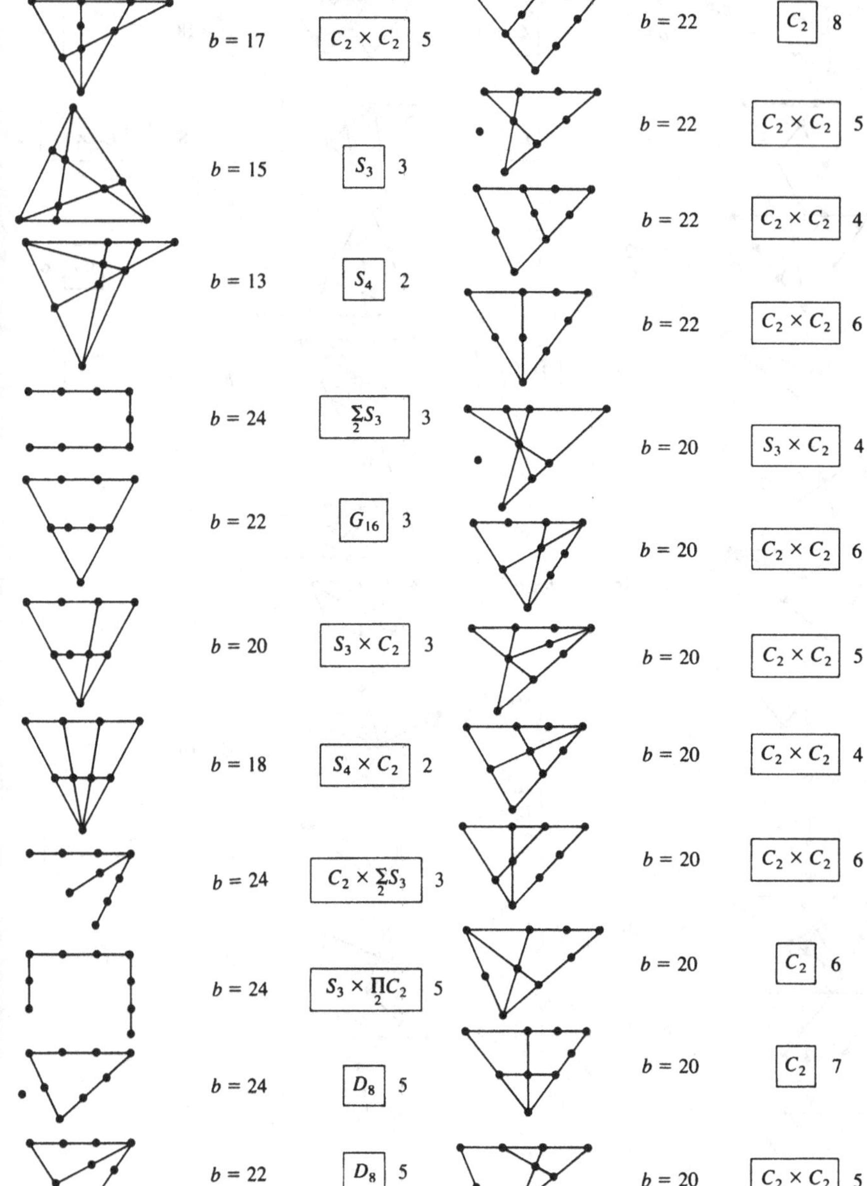
$b = 17$ $C_2 \times C_2$ 5
$b = 15$ S_3 3
$b = 13$ S_4 2
$b = 24$ $\sum_2 S_3$ 3
$b = 22$ G_{16} 3
$b = 20$ $S_3 \times C_2$ 3
$b = 18$ $S_4 \times C_2$ 2
$b = 24$ $C_2 \times \sum_2 S_3$ 3
$b = 24$ $S_3 \times \prod_2 C_2$ 5
$b = 24$ D_8 5
$b = 22$ D_8 5
$b = 22$ C_2 8
$b = 22$ $C_2 \times C_2$ 5
$b = 22$ $C_2 \times C_2$ 4
$b = 22$ $C_2 \times C_2$ 6
$b = 20$ $S_3 \times C_2$ 4
$b = 20$ $C_2 \times C_2$ 6
$b = 20$ $C_2 \times C_2$ 5
$b = 20$ $C_2 \times C_2$ 4
$b = 20$ $C_2 \times C_2$ 6
$b = 20$ C_2 6
$b = 20$ C_2 7
$b = 20$ $C_2 \times C_2$ 5

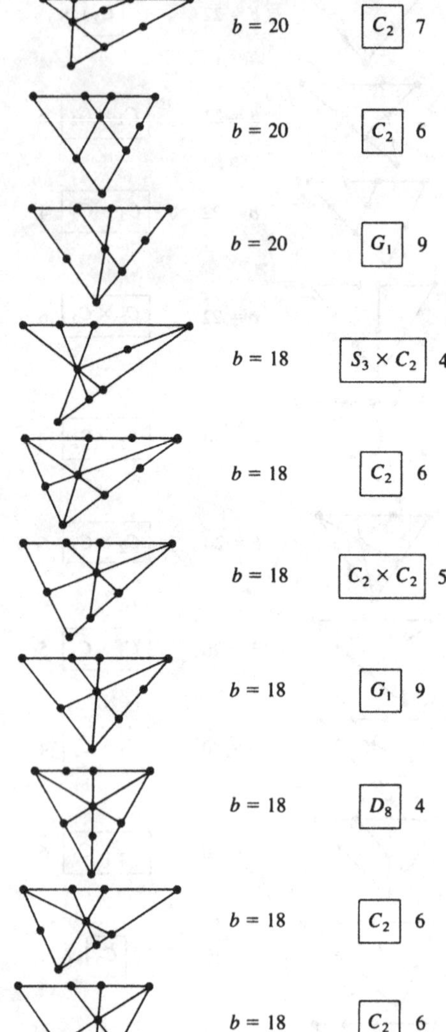

b	Group	
$b = 20$	C_2	7
$b = 20$	C_2	6
$b = 20$	G_1	9
$b = 18$	$S_3 \times C_2$	4
$b = 18$	C_2	6
$b = 18$	$C_2 \times C_2$	5
$b = 18$	G_1	9
$b = 18$	D_8	4
$b = 18$	C_2	6
$b = 18$	C_2	6
$b = 18$	G_1	9

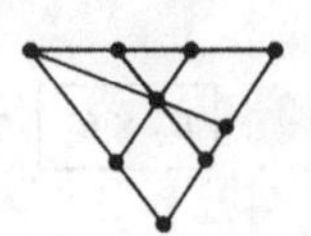

$b = 18$ G_1 9

$b = 18$ $C_2 \times C_2$ 4

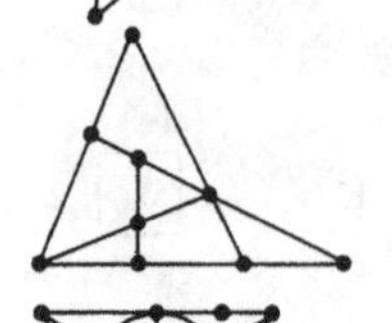

$b = 18$ C_2 6

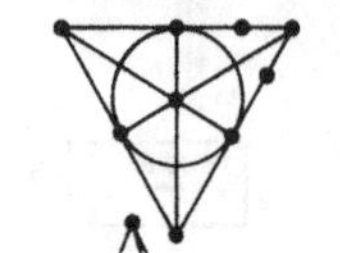

$b = 16$ D_8 4

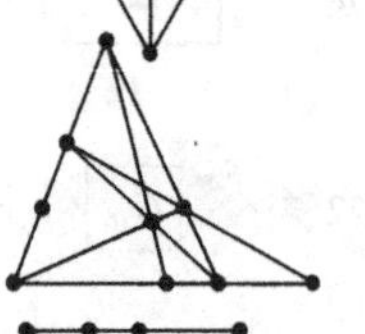

$b = 16$ C_2 6

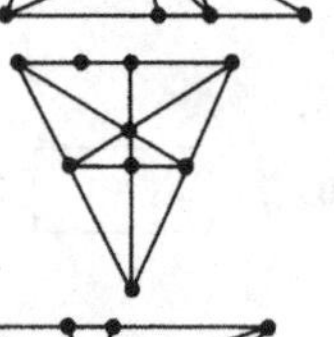

$b = 16$ $C_2 \times C_2$ 6

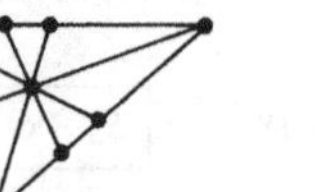

$b = 16$ C_2 6

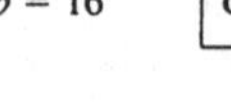

$b = 16$ $C_2 \times C_2$ 4

$b = 16$ C_2 6

$b = 16$ C_2 6

$b = 16$ C_2 6

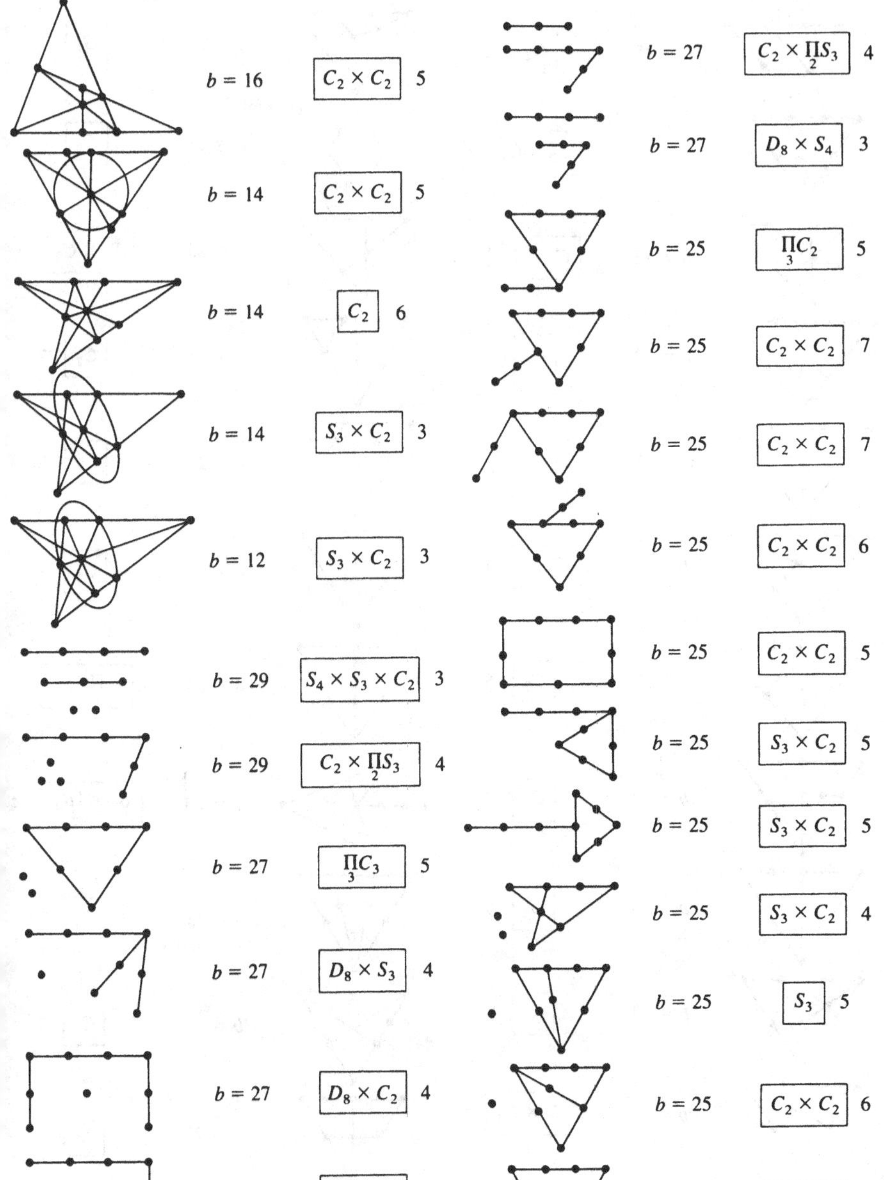

$b = 16$ $C_2 \times C_2$ 5
$b = 14$ $C_2 \times C_2$ 5
$b = 14$ C_2 6
$b = 14$ $S_3 \times C_2$ 3
$b = 12$ $S_3 \times C_2$ 3
$b = 29$ $S_4 \times S_3 \times C_2$ 3
$b = 29$ $C_2 \times \prod_2 S_3$ 4
$b = 27$ $\prod_3 C_3$ 5
$b = 27$ $D_8 \times S_3$ 4
$b = 27$ $D_8 \times C_2$ 4
$b = 27$ $S_3 \times C_2$ 6
$b = 27$ $C_2 \times \prod_2 S_3$ 4
$b = 27$ $D_8 \times S_4$ 3
$b = 25$ $\prod_3 C_2$ 5
$b = 25$ $C_2 \times C_2$ 7
$b = 25$ $C_2 \times C_2$ 7
$b = 25$ $C_2 \times C_2$ 6
$b = 25$ $C_2 \times C_2$ 5
$b = 25$ $S_3 \times C_2$ 5
$b = 25$ $S_3 \times C_2$ 5
$b = 25$ $S_3 \times C_2$ 4
$b = 25$ S_3 5
$b = 25$ $C_2 \times C_2$ 6
$b = 25$ $C_2 \times C_2$ 6

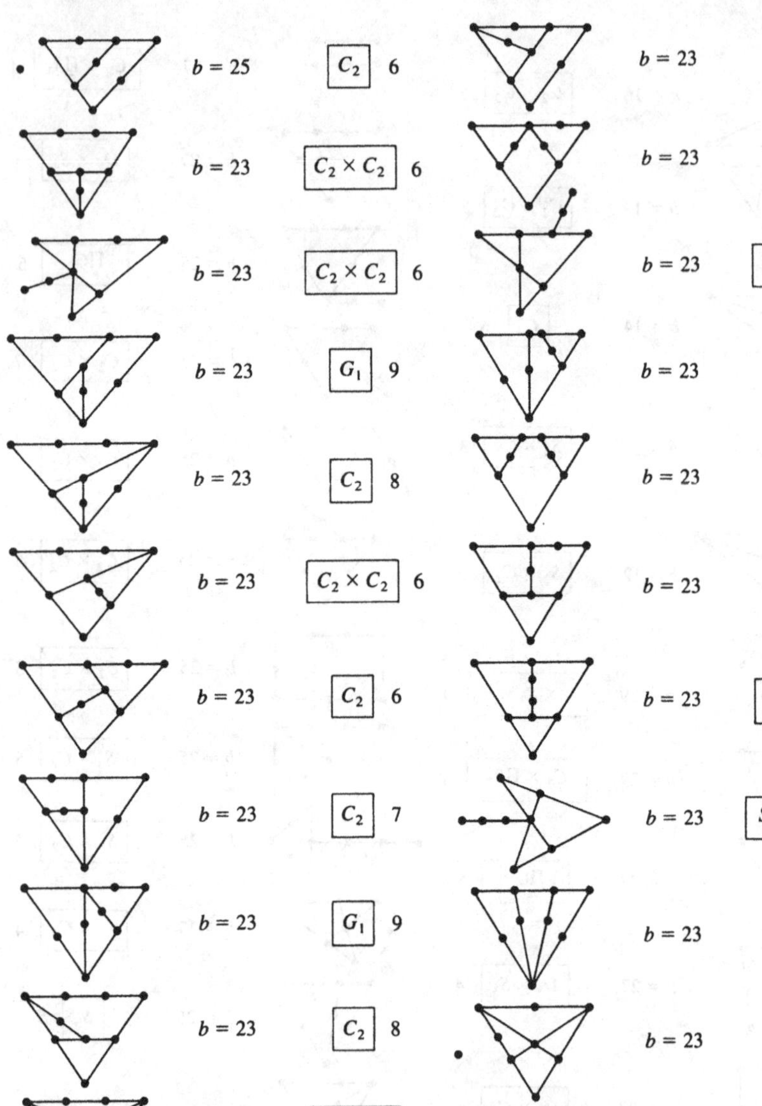
$b = 25$ C_2 6
$b = 23$ $C_2 \times C_2$ 6
$b = 23$ $C_2 \times C_2$ 6
$b = 23$ G_1 9
$b = 23$ C_2 8
$b = 23$ $C_2 \times C_2$ 6
$b = 23$ C_2 6
$b = 23$ C_2 7
$b = 23$ G_1 9
$b = 23$ C_2 8
$b = 23$ $C_2 \times C_2$ 5
$b = 23$ $C_2 \times C_2$ 6
$b = 23$ G_1 9
$b = 23$ C_2 6
$b = 23$ $S_3 \times C_2$ 4
$b = 23$ C_2 7
$b = 23$ C_2 5
$b = 23$ C_2 7
$b = 23$ $\prod_3 C_2$ 4
$b = 23$ $S_3 \times \prod_2 C_2$ 4
$b = 23$ S_4 3
$b = 23$ C_2 7
$b = 23$ S_3 5
$b = 23$ C_2 7

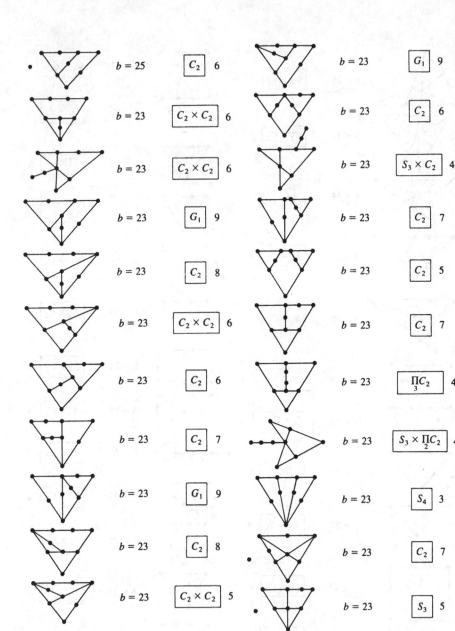

• 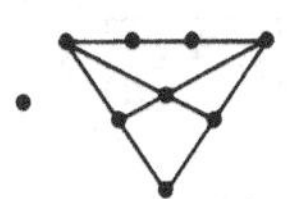 $b = 23$ $\boxed{D_8 \times C_2}$ 4

• $b = 23$ $\boxed{C_2 \times C_2}$ 5

• $b = 23$ $\boxed{D_8}$ 3

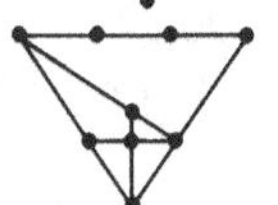 $b = 21$ $\boxed{C_2 \times C_2}$ 6

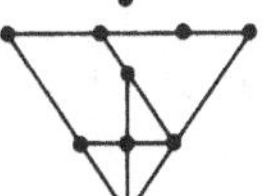 $b = 21$ $\boxed{C_2}$ 6

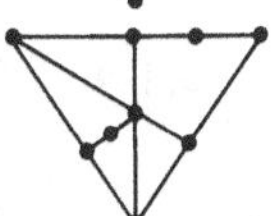 $b = 21$ $\boxed{G_1}$ 9

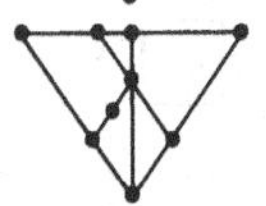 $b = 21$ $\boxed{G_1}$ 9

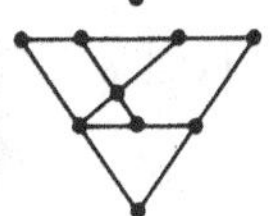 $b = 21$ $\boxed{C_2}$ 5

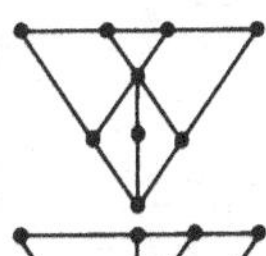 $b = 21$ $\boxed{C_2 \times C_2}$ 4

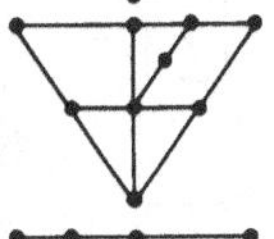 $b = 21$ $\boxed{C_2}$ 7

$b = 21$ $\boxed{C_2}$ 6

$b = 21$ $\boxed{G_1}$ 9

$b = 21$ $\boxed{C_2}$ 6

$b = 21$ $\boxed{G_1}$ 9

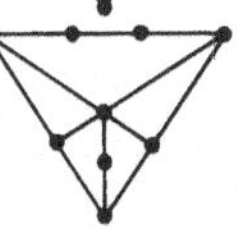 $b = 21$ $\boxed{\prod_3 C_2}$ 5

$b = 21$ $\boxed{C_2 \times C_2}$ 5

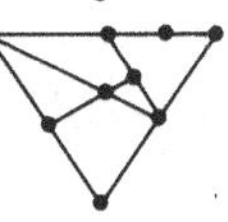 $b = 21$ $\boxed{G_1}$ 9

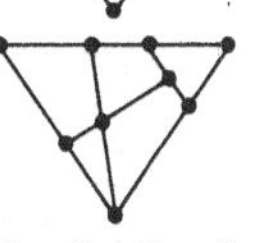 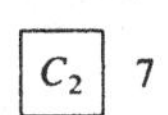 $b = 21$ $\boxed{C_2}$ 7

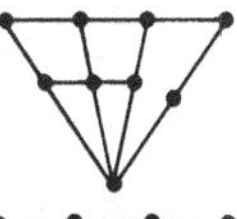 $b = 21$ $\boxed{S_3}$ 5

 $b = 21$ $\boxed{C_2}$ 7

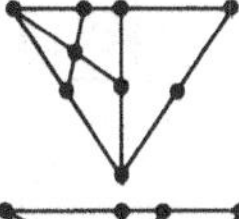 $b = 21$ $\boxed{G_1}$ 9

 $b = 21$ $\boxed{G_1}$ 9

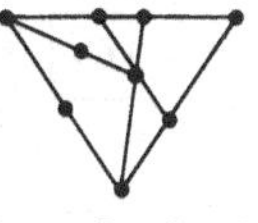 $b = 21$ $\boxed{C_2}$ 6

$b = 21$ $\boxed{G_1}$ 9

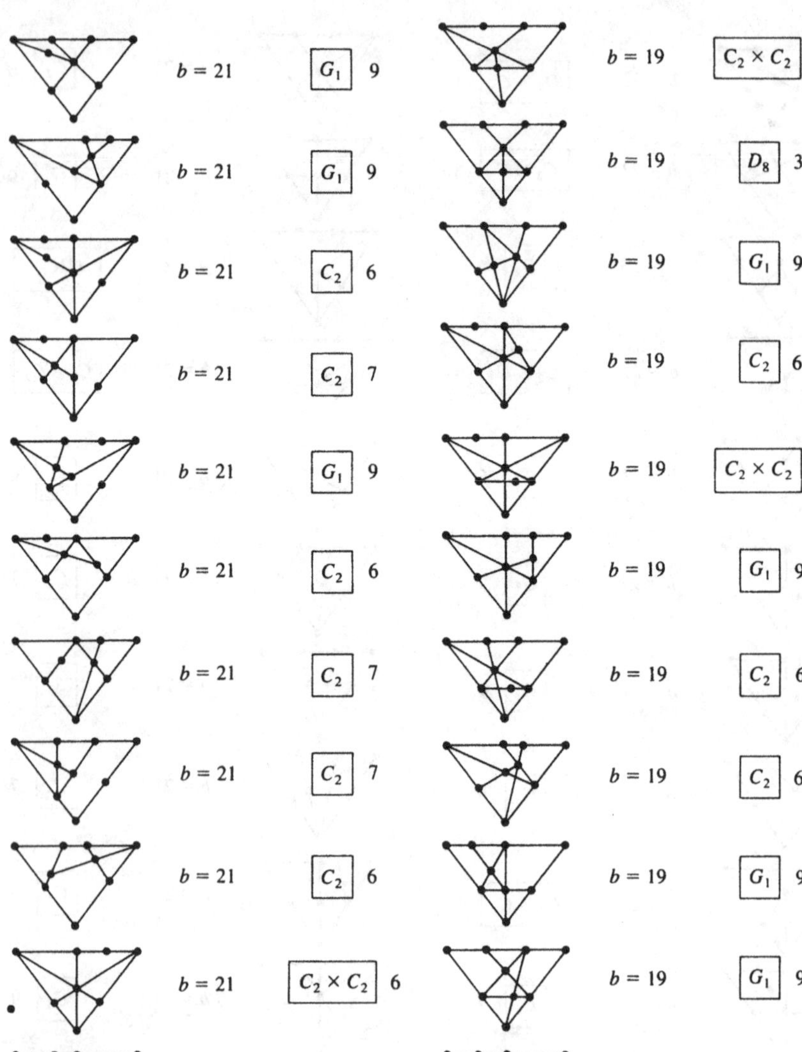
$b = 21$ G_1 9
$b = 21$ G_1 9
$b = 21$ C_2 6
$b = 21$ C_2 7
$b = 21$ G_1 9
$b = 21$ C_2 6
$b = 21$ C_2 7
$b = 21$ C_2 7
$b = 21$ C_2 6
$b = 21$ $C_2 \times C_2$ 6
$b = 21$ C_2 6
$b = 19$ $D_8 \times C_2$ 4
$b = 19$ $C_2 \times C_2$ 5
$b = 19$ D_8 3
$b = 19$ G_1 9
$b = 19$ C_2 6
$b = 19$ $C_2 \times C_2$ 6
$b = 19$ G_1 9
$b = 19$ C_2 6
$b = 19$ C_2 6
$b = 19$ G_1 9
$b = 19$ G_1 9
$b = 19$ G_1 9
$b = 19$ C_2 7

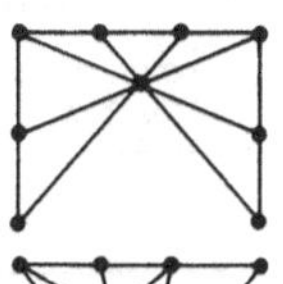
$b = 19$ $\boxed{C_2}$ 5

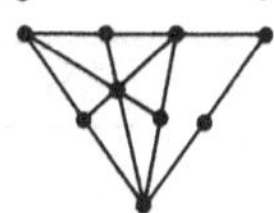
$b = 19$ $\boxed{C_2}$ 7

$b = 19$ $\boxed{G_1}$ 9

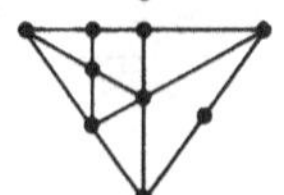
$b = 19$ $\boxed{G_1}$ 9

$b = 19$ $\boxed{G_1}$ 9

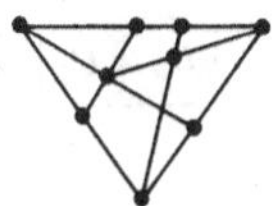
$b = 19$ $\boxed{C_2}$ 5

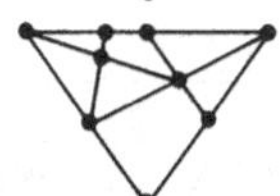
$b = 19$ $\boxed{G_1}$ 9

$b = 19$ $\boxed{G_1}$ 9

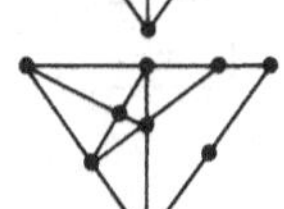
$b = 19$ $\boxed{C_2}$ 7

$b = 19$ $\boxed{G_1}$ 9

$b = 19$ $\boxed{C_2}$ 6

$b = 19$ $\boxed{C_2}$ 6

$b = 19$ $\boxed{S_3}$ 4

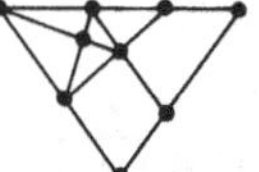
$b = 19$ $\boxed{C_2}$ 6

$b = 19$ $\boxed{C_2 \times C_2}$ 5

$b = 19$ $\boxed{C_2}$ 7

$b = 19$ $\boxed{S_4}$ 4

$b = 19$ $\boxed{D_8}$ 4

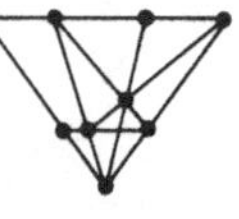
$b = 17$ $\boxed{C_2}$ 7

$b = 17$ $\boxed{G_1}$ 9

$b = 17$ $\boxed{C_2}$ 5

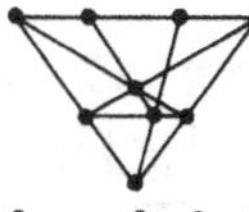
$b = 17$ $\boxed{C_2 \times C_2}$ 5

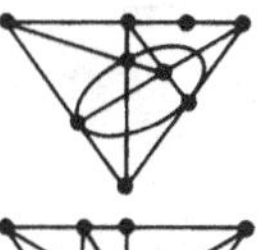
$b = 17$ $\boxed{S_3}$ 4

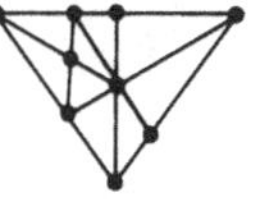
$b = 17$ $\boxed{G_1}$ 9

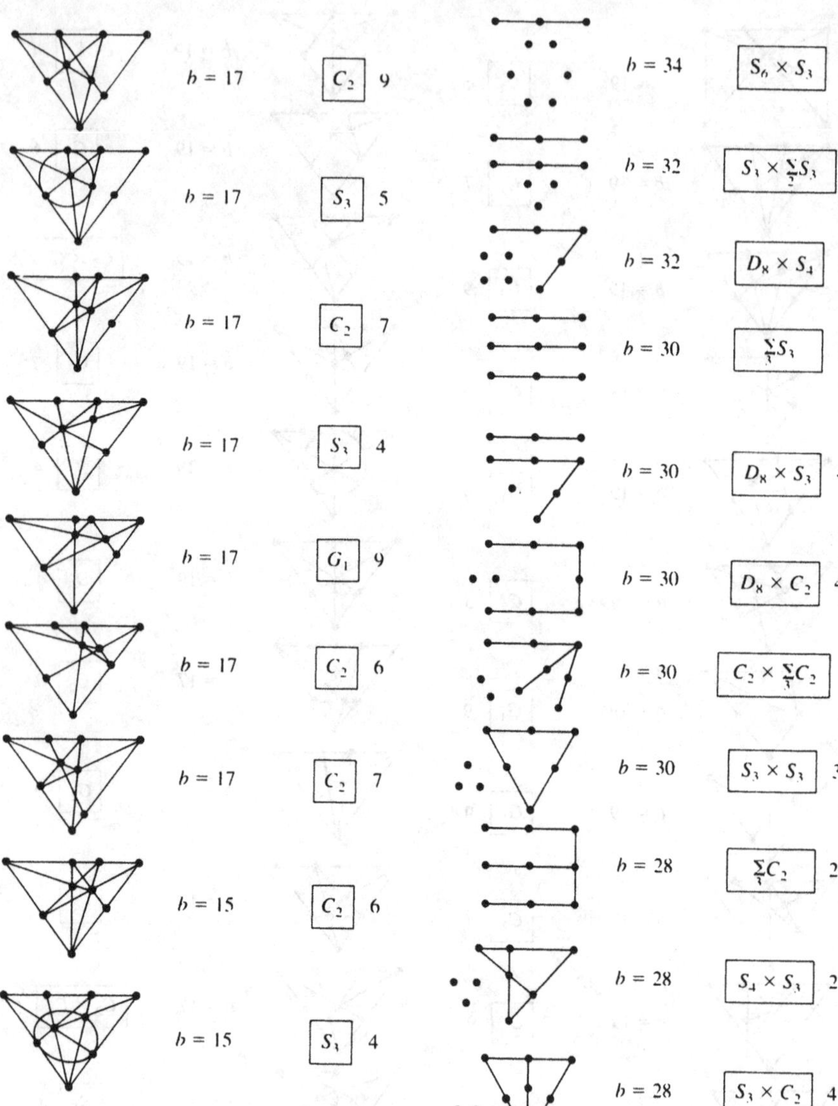
b = 17
C₂ 9
b = 17
S₃ 5
b = 17
C₂ 7
b = 17
S₃ 4
b = 17
G₁ 9
b = 17
C₂ 6
b = 17
C₂ 7
b = 15
C₂ 6
b = 15
S₃ 4
b = 15
C₄ 3
b = 34
S₆ × S₃ 2
b = 32
S₃ × Σ₂S₃ 2
b = 32
D₈ × S₄ 3
b = 30
Σ₃S₃ 1
b = 30
D₈ × S₃ 4
b = 30
D₈ × C₂ 4
b = 30
C₂ × Σ₃C₂ 3
b = 30
S₃ × S₃ 3
b = 28
Σ₃C₂ 2
b = 28
S₄ × S₃ 2
b = 28
S₃ × C₂ 4
b = 28
∐₃C₂ 4

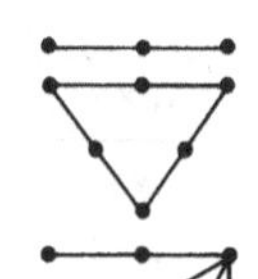

$b = 28$ $\boxed{S_3 \times S_3}$ 3

$b = 28$ $\boxed{\sum_4 C_2}$ 2

$b = 28$ $\boxed{D_8}$ 3

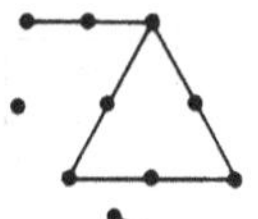

$b = 28$ $\boxed{C_2 \times C_2}$ 6

$b = 28$ $\boxed{C_2 \times C_2}$ 6

$b = 28$ $\boxed{D_8}$ 4

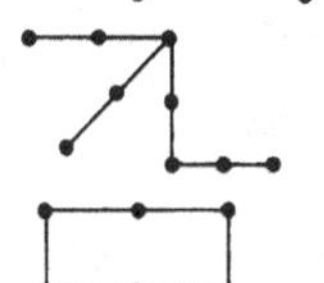

$b = 28$ $\boxed{D_8 \times C_2}$ 5

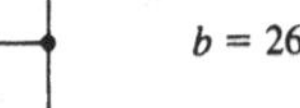

$b = 26$ $\boxed{S_3 \times C_2}$ 2

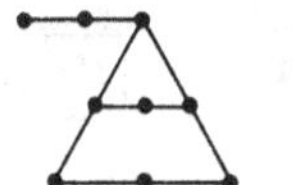

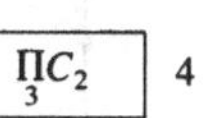

$b = 26$ $\boxed{\prod_3 C_2}$ 4

$b = 26$ $\boxed{S_4 \times S_3}$ 2

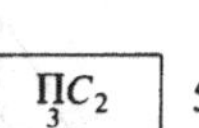

$b = 26$ $\boxed{\prod_3 C_2}$ 5

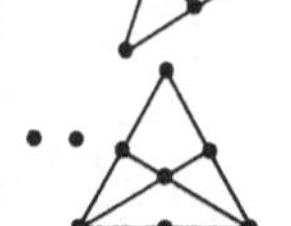

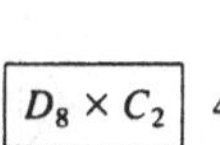

$b = 26,$ $\boxed{D_8 \times C_2}$ 4

$b = 26$ $\boxed{S_3 \times C_2}$ 4

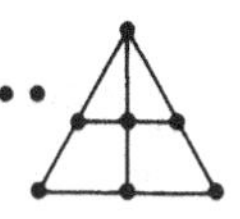

$b = 26$ $\boxed{S_3 \times \prod_2 C_2}$ 3

$b = 26$ $\boxed{C_2}$ 7

$b = 26$ $\boxed{C_2}$ 6

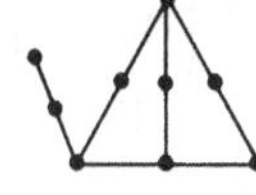

$b = 26$ $\boxed{C_2 \times C_2}$ 6

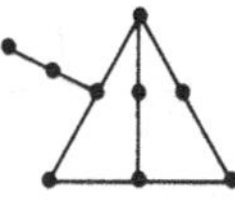

$b = 26$ $\boxed{C_2 \times C_2}$ 6

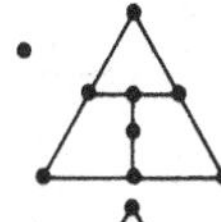

$b = 26$ $\boxed{C_2 \times C_2}$ 5

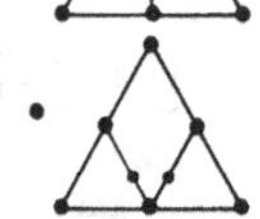

$b = 26$ $\boxed{C_2}$ 6

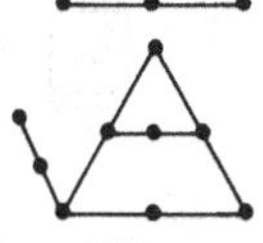

$b = 26$ $\boxed{C_2}$ 8

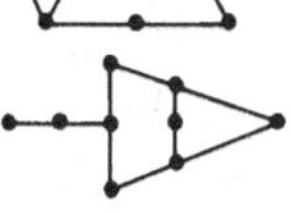

$b = 26$ $\boxed{C_2 \times C_2}$ 6

$b = 26$ $\boxed{C_2 \times C_2}$ 4

$b = 26$ $\boxed{C_2}$ 6

$b = 26$ $\boxed{C_2}$ 7

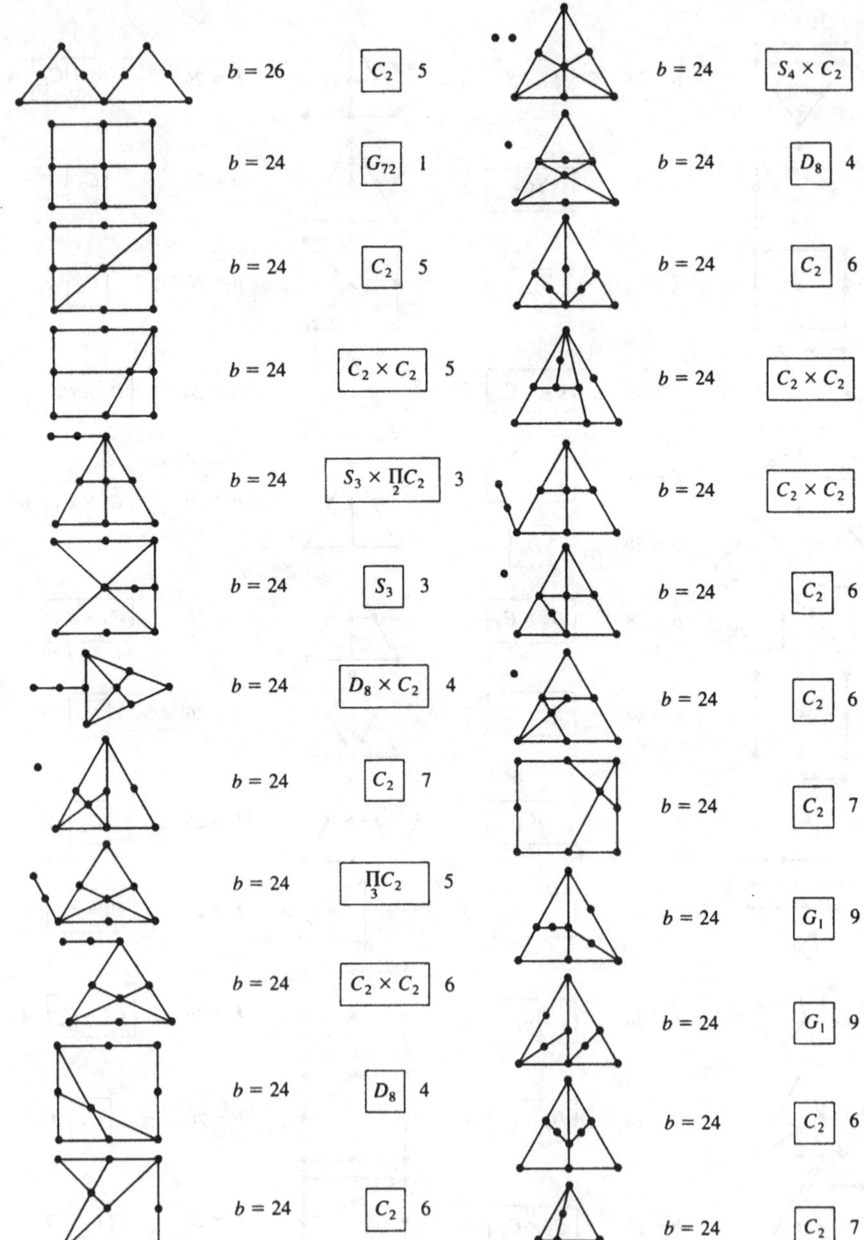

b = 26
C_2
5
b = 24
G_{72}
1
b = 24
C_2
5
b = 24
$C_2 \times C_2$
5
b = 24
$S_3 \times \prod_2 C_2$
3
b = 24
S_3
3
b = 24
$D_8 \times C_2$
4
b = 24
C_2
7
b = 24
$\prod_3 C_2$
5
b = 24
$C_2 \times C_2$
6
b = 24
D_8
4
b = 24
C_2
6
b = 24
$S_4 \times C_2$
3
b = 24
D_8
4
b = 24
C_2
6
b = 24
$C_2 \times C_2$
4
b = 24
$C_2 \times C_2$
6
b = 24
C_2
6
b = 24
C_2
6
b = 24
C_2
7
b = 24
G_1
9
b = 24
G_1
9
b = 24
C_2
6
b = 24
C_2
7

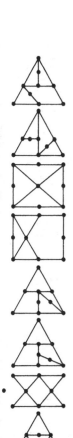

$b = 24$ $\boxed{G_1}$ 9

$b = 24$ $\boxed{G_1}$ 9

$b = 24$ $\boxed{D_8}$ 3

$b = 24$ $\boxed{C_2}$ 6

$b = 24$ $\boxed{G_1}$ 9

$b = 24$ $\boxed{G_1}$ 9

• $b = 24$ $\boxed{S_3 \times C_2}$ 3

$b = 24$ $\boxed{S_3}$ 3

$b = 24$ $\boxed{C_2}$ 6

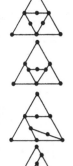

$b = 24$ $\boxed{C_2}$ 5

$b = 24$ $\boxed{S_3 \times C_2}$ 2

$b = 22$ $\boxed{S_3 \times C_2}$ 2

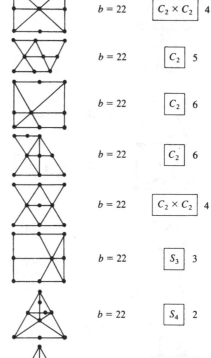

$b = 22$ $\boxed{C_2 \times C_2}$ 4

$b = 22$ $\boxed{C_2}$ 5

$b = 22$ $\boxed{C_2}$ 6

$b = 22$ $\boxed{C_2}$ 6

$b = 22$ $\boxed{C_2 \times C_2}$ 4

$b = 22$ $\boxed{S_3}$ 3

$b = 22$ $\boxed{S_4}$ 2

$b = 22$ $\boxed{D_8 \times C_2}$ 4

$b = 22$ $\boxed{D_8}$ 4

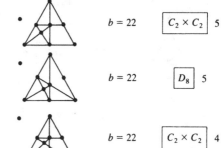

• $b = 22$ $\boxed{C_2 \times C_2}$ 5

• $b = 22$ $\boxed{D_8}$ 5

• $b = 22$ $\boxed{C_2 \times C_2}$ 4

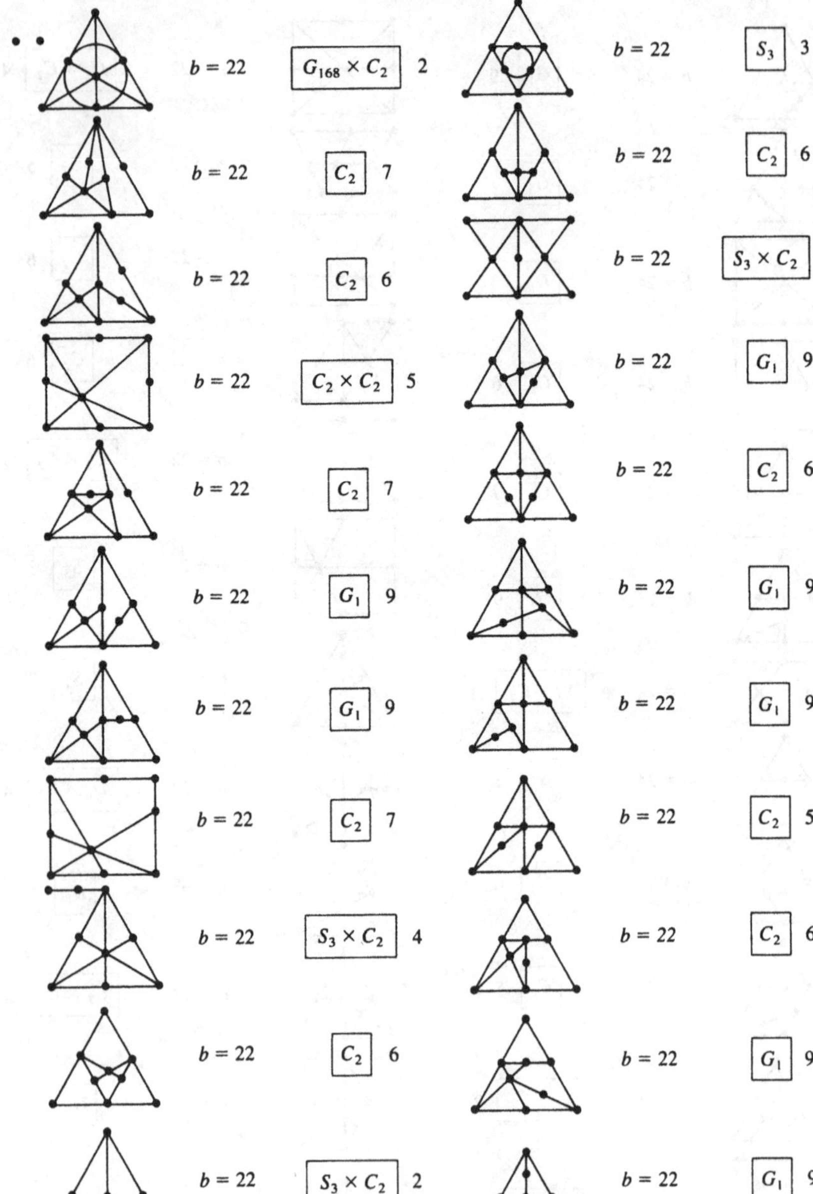

b = 22 $G_{168} \times C_2$ 2
b = 22 C_2 7
b = 22 C_2 6
b = 22 $C_2 \times C_2$ 5
b = 22 C_2 7
b = 22 G_1 9
b = 22 G_1 9
b = 22 C_2 7
b = 22 $S_3 \times C_2$ 4
b = 22 C_2 6
b = 22 $S_3 \times C_2$ 2
b = 22 S_3 3
b = 22 C_2 6
b = 22 $S_3 \times C_2$ 3
b = 22 G_1 9
b = 22 C_2 6
b = 22 G_1 9
b = 22 G_1 9
b = 22 C_2 5
b = 22 C_2 6
b = 22 G_1 9
b = 22 G_1 9

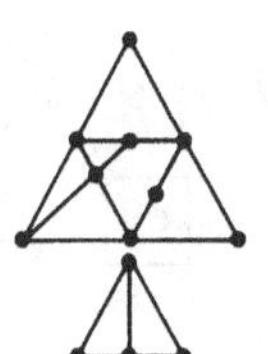

$b = 22$ $\boxed{G_1}$ 9

$b = 22$ $\boxed{G_1}$ 9

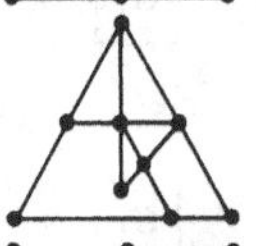

$b = 22$ $\boxed{G_1}$ 9

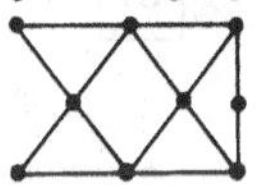

$b = 22$ $\boxed{C_2}$ 6

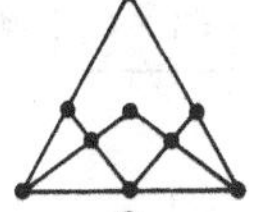

$b = 22$ $\boxed{C_2 \times C_2}$ 4

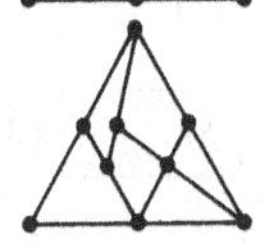

$b = 22$ $\boxed{G_1}$ 9

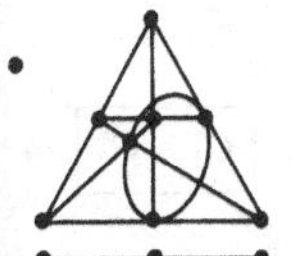

$b = 20$ $\boxed{G_{48}}$ 2

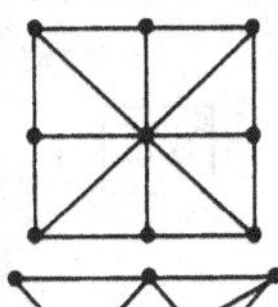

$b = 20$ $\boxed{D_8}$ 3

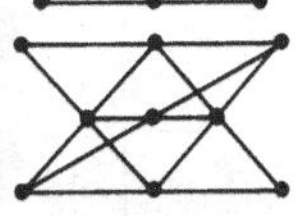

$b = 20$ $\boxed{S_3 \times C_2}$ 2

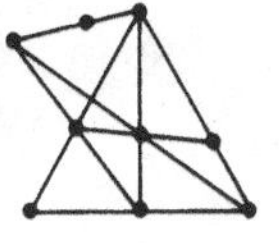

$b = 20$ $\boxed{C_2}$ 6

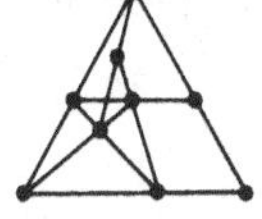

$b = 20$ $\boxed{C_2}$ 6

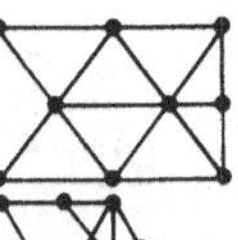

$b = 20$ $\boxed{C_2}$ 6

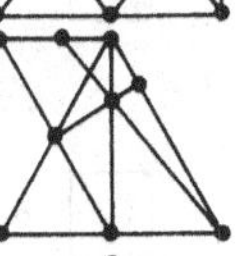

$b = 20$ $\boxed{C_2}$ 5

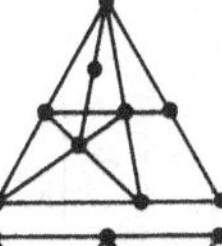

$b = 20$ $\boxed{C_2 \times C_2}$ 5

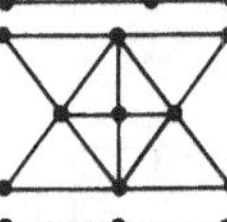

$b = 20$ $\boxed{C_2 \times C_2}$ 4

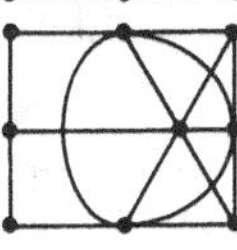

$b = 20$ $\boxed{S_4}$ 2

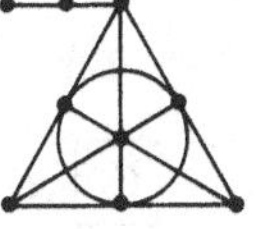

$b = 20$ $\boxed{S_4 \times C_2}$ 3

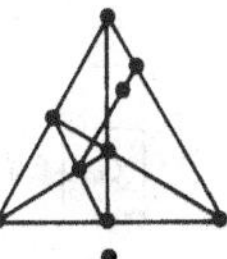

$b = 20$ $\boxed{D_8}$ 5

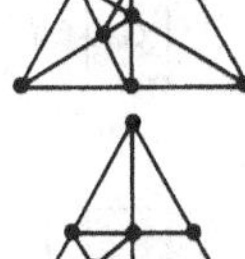

$b = 20$ $\boxed{C_2}$ 7

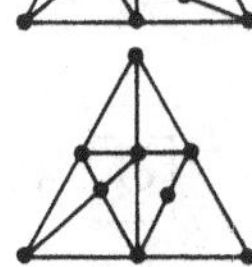

$b = 20$ $\boxed{G_1}$ 9

$b = 20$ $\boxed{D_8}$ 4

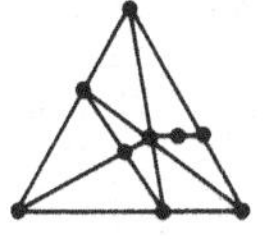

$b = 20$ $\boxed{C_2}$ 7

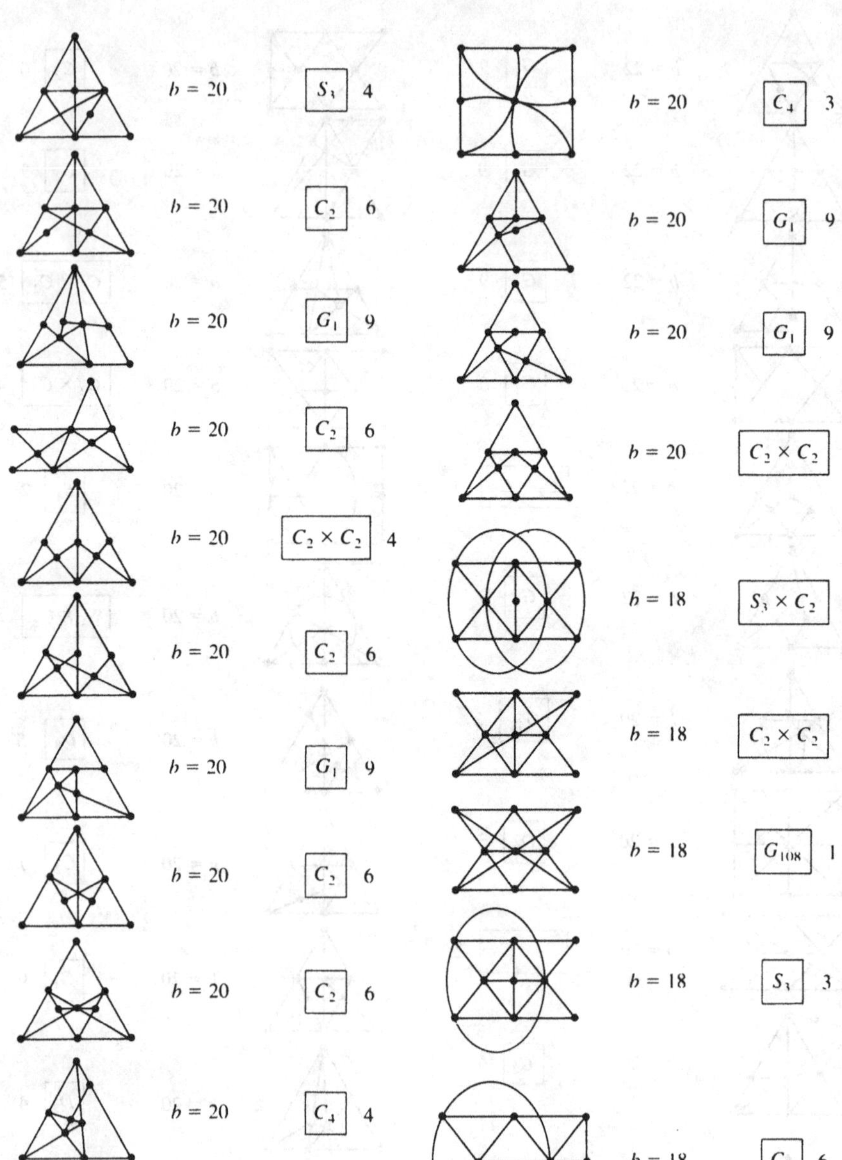

$b = 20$ S_3 4
$b = 20$ C_2 6
$b = 20$ G_1 9
$b = 20$ C_2 6
$b = 20$ $C_2 \times C_2$ 4
$b = 20$ C_2 6
$b = 20$ G_1 9
$b = 20$ C_2 6
$b = 20$ C_2 6
$b = 20$ C_4 4
$b = 20$ G_1 9
$b = 20$ C_4 3
$b = 20$ G_1 9
$b = 20$ G_1 9
$b = 20$ $C_2 \times C_2$ 4
$b = 18$ $S_3 \times C_2$ 3
$b = 18$ $C_2 \times C_2$ 4
$b = 18$ G_{108} 1
$b = 18$ S_3 3
$b = 18$ C_2 6

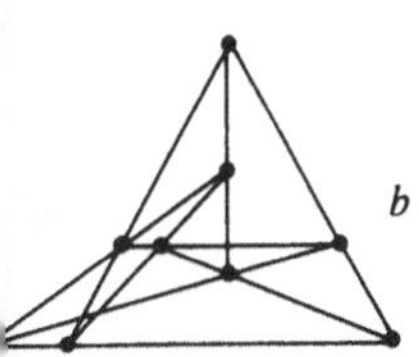

$b = 18$ $\boxed{S_3 \times C_2}$ 2

$b = 18$ $\boxed{C_2}$ 6

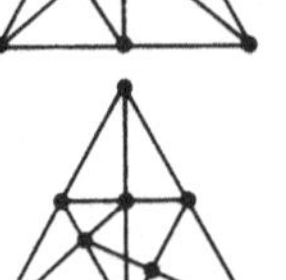

$b = 18$ $\boxed{G_1}$ 9

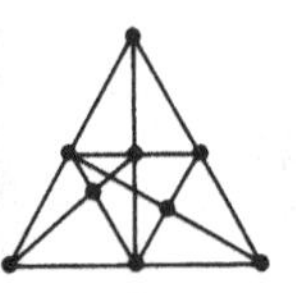

$b = 18$ $\boxed{C_2}$ 6

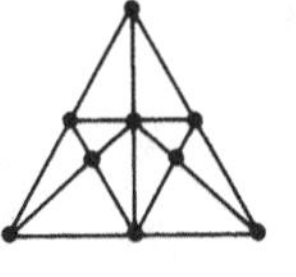

$b = 18$ $\boxed{C_2 \times C_2}$ 4

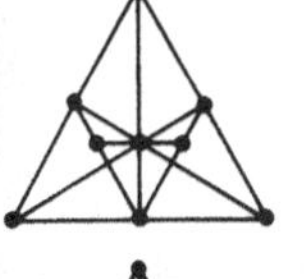

$b = 18$ $\boxed{C_2}$ 6

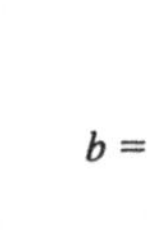

$b = 18$ $\boxed{C_9}$ 1

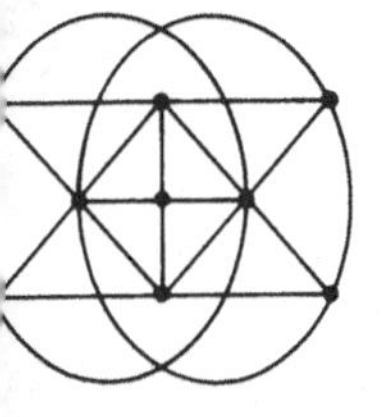

$b = 16$ $\boxed{D_8}$ 3

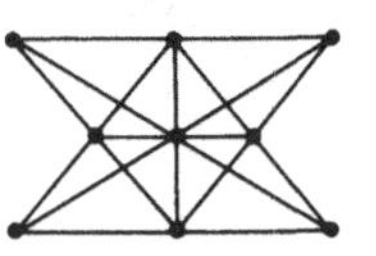

$b = 16$ $\boxed{G_{36}}$ 2

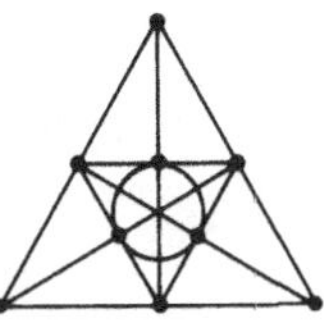

$b = 16$ $\boxed{S_3}$ 3

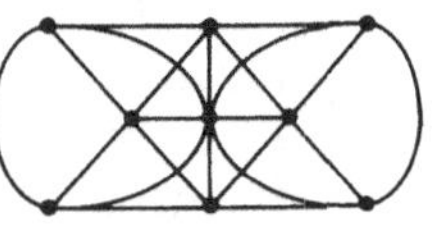

$b = 16$ $\boxed{S_3 \times C_2}$ 2

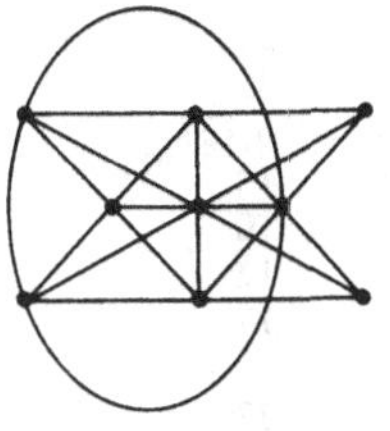

$b = 14$ $\boxed{G_{36}}$ 2

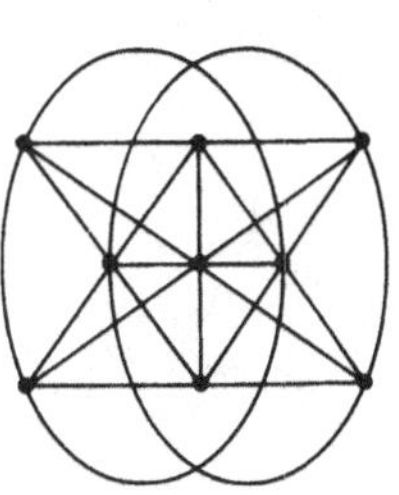

$b = 12$ $\boxed{G_{432}}$ 1

Notation index

pq	1
v	1
b	1
r_p	2
k_L	2
K_v	2
$(k_1, \ldots, k_s)$-star	2
(k_1, k_2)-cross	3
$\langle X \rangle$	3, 7
$\parallel$	6,139
$AG(d, n)$	10
$PG(d, n)$	10
$[a]$	12
t-(v, k, λ) design	17
$S(2, 3, v)$	18
$(r, 1)$-design	23
k-arc	28
b_k	33
v_r	33
$\bar{s}$	49
$\pi(p, L)$	68
$\{0, s, t\}$-semiaffine	69
$S_{s,t}$	72
$m(L, H)$	96
$\langle X_1, \ldots, X_s \rangle$	137
v_V	137
b_V	137
v_i	137
b_i	137
r'_p	137
Aut(S)	169
v_G	170
b_G	170
G_L	175

Subject index

affine plane, 3
affine space, 6
antiflag, 169
arc, 28
automorphism group, 168

Baer subplane, 30
basis, 9
biaffine plane, 68
Bundle Theorem, 149

chain
  4-chain, 178
characterization result, 12
claw, 97
  normal, 97
clique, 97
  maximal, 97
complete graph, 2
cone, 153
Conway, 16

degenerate, 7
degree, 2
Dembowski order, 13
Desarguesian space, 10
design
  t-(v, k, λ) design, 17
  (r, l)-design, 23
dimension, 9
  d-dimensional, 9, 136
dual, 8

embedded, 11
extended Nwankpa–Shrikhande
  plane, 19
extension, 23
  s times extendible, 23

Fano plane, 8
Fano quasi-plane, 92, 119
flag, 169
Fundamental Theorem, 15

generalized projective, 9

Hering affine plane, 178
Hermitian unital, 176
homogeneous, 172, 173
hyperideal line, 126
hyperoval, 28
hyperplane, 9

i-line, 2
i-point, 2
i-space, 137
i-subspace, 137
l-affine, 68
l-semiaffine, 68
ideal line, point, 126
inversive plane, 178
inversive space, 152
isomorphic, 20

k-arc, 28
knot, 153

linear space, 1
  at infinity, 119
  d-dimensional, 136
  finite, 1
  restricted, 119
  trivial, 1
locally projective, 146

Miquelian, 178

nearfield, 177
nearfield affine plane, 177
near-pencil, 2

Netto system, 178
normal, 97
nucleus space, 153
Nwankpa plane, 19
Nwankpa–Shrikhande plane, 18

orbit, 169
  line, 169
  point, 169
order,
  of affine plane, 5
  of affine space, 6
  of claw, 97
  of linear space, 13
  of projective plane, 9
  of projective space, 10

parallel, 6
parallel class, 5
parallelism, 6
parameter, 2
partial plane, 95
pencil
  a-pencil, 34
plane, 9, 138
plane *k*-arc, 153
point at infinity, 10
projective plane, 7
  degenerative, 7
projective space, 9
pseudo-complement, 22
π-space, 138
punctured, 19

quadric, 153
  elliptic, 153
  hyperbolic, 153
  parabolic, 153

real line, point, 126
Ree group, 176
Ree unital, 177
restricted, 119

semiaffine plane, 68
semiaffinity condition, 12
space
  i-space, 137
  π-space, 138
square order, 13
stabilizer, 175
star, 2
Steiner triple system, 18
subplane, 30
subspace, 9, 136
  linear subspace, 9
transitive, 169, 172, 173
  antiflag, 169
  flag, 169
  line, 169
  point, 169
triangle, 27

unital, 31, 176
  Hermitian, 176
  Ree, 177

Veblen–Young axiom, 151